And others

Handbook of Plant Dissection

And others

Handbook of Plant Dissection

ISBN/EAN: 9783337250157

Printed in Europe, USA, Canada, Australia, Japan

Cover: Foto ©berggeist007 / pixelio.de

More available books at **www.hansebooks.com**

HANDBOOK

OF

PLANT DISSECTION

BY

J. C. ARTHUR, M.Sc.,

Botanist to the New York Agricultural Experiment-Station,

CHARLES R. BARNES, M.A.,

Professor of Botany in Purdue University,

AND

JOHN M. COULTER, Ph.D.,

Professor of Botany in Wabash College,

EDITORS OF THE

BOTANICAL GAZETTE.

NEW YORK

HENRY HOLT AND COMPANY

1886

Press of W. L. Mershon & Co.,
Rahway, N . J,

PREFACE.

A rich harvest of laboratory manuals has resulted to zoölogy from the publication of Huxley and Martin's Elementary Biology ten years ago. Although that work embraced both animals and plants with over half the examples from the latter, it has given rise to no similar aid to botanical study till the past year. The increasing laboratory facilities in this country seem to warrant the expectation that an elementary manual like the present work will now be found in many instances to afford welcome assistance to both teacher and pupil.

In 1882 one of the authors of this book drew up an outline of work for a few plants, which was used in the Summer School of Science of the University of Minnesota. Not long afterward the preparation of the present hand-book was actively undertaken by the three authors conjointly, and has since been gradually perfected and tested by repeated use with classes and individual students.

Although the present work is based upon Huxley and Martin's in form and mode of treatment for the laboratory part, it differs in excluding all matters of physiology so far as possible, as the present demands of vegetable physiology will hardly permit harmonious treatment along with a course of dissection.

In drawing up the outlines of work the aim has been to direct the student in a very careful and systematic examination of a few examples, so that while he is securing a knowledge of the main features of plant anatomy, he will at the same time acquire the habit of close and critical observation, which is indispensable to the successful prosecution of natural history studies. To this end the directions for finding the different parts have been made as explicit as possible, and at the same time as little information given about them as seemed advisable ; for the student having found the part is expected to examine it thoroughly until he has found out all that may be readily seen. This rule has been modified according to the difficulties to be overcome, and in extreme cases full information has been provided, which the student is only expected to verify. On the other hand, it will repeatedly happen that more may be learned by an acute observer than there is any hint of in the outlines, as the work, though deemed sufficiently exhaustive for the student, is far from being so for the specialist.

In the use of such outlines as these there is always danger that the student will slight the study of those parts which he is expected to work out for himself and only attempt to verify the portions where the information is fuller. If it be found that too great dependence is being placed on the manual it will be advisable to substitute plants allied to those named, thus withdrawing all exact information ; the laboratory directions will still serve as a guide to the order and methods of examination.

It has been no part of the present aim to provide a key to the nomenclature of plant anatomy. When technical terms are used, as indeed is necessarily very frequent, they have usually been preceded by descriptive definitions, either direct or implied. A glossary is added to further assist the student, so that he may find as little difficulty with the names as possible, and devote himself chiefly to the objects themselves. On this account, and on account of the progressive series of forms which have been chosen, it is hoped that the **work will be** found suitable **not** only for classes pursuing a regular course of lectures, but also for those who have never before studied **botany,** and **for** home use away **from** the assistance **of a** teacher.

The required **apparatus, reagents and materials have been** reduced **to a minimum,** difficult manipulations (except the **cutting of sufficiently** thin sections) have, **to** a large extent, **been** excluded, and the minute anatomy has been kept within the limits of the average microscope used **in** the botanical laboratories of this country. In short, the attempt has been to provide a guide to the study of a few common plants in **which** simple appliances, coupled with persever-ance **and** keen observation on **the part of the** learner, are the only essentials.

Under "gross anatomy" the plant is first examined with the **aid** only **of a hand lens,** and **then** passing to "minute anatomy," every **part is subjected to** the compound **micro-**scope. **A student's** success in the latter **may often be** gauged by his ability **to** discover **all** there **is to be seen** under the former.

The laboratory work for each plant is preceded by direc-**tions for the** preliminary finding and preparation of mate-**rial. It** is followed by annotations which serve a number of purposes : (1) to explain obscure matters, (2) to give additional information which for want of higher powers, special reagents **or proper** materials, the student **is** unable in the usual limited **time** to secure for himself, but which is essential to fully round **out the** subject, more especially, however, (3) to give some insight into the course of develop-**ment from** the lower to **the** higher forms which will serve as a thread on which the **most** important facts ascertained in the laboratory **work may be** strung, **and not** the least (4) **to direct** the student to sources of additional information by means of **which he may pursue his inquiries as far as he** may choose. The annotations are necessarily fragmentary and disconnected, and the references to literature only sufficient to start the student in his researches.

January, 1886.   THE AUTHORS.

CONTENTS.

EXPLANATION OF PLATE I.

ILLUSTRATIONS IN GROSS ANATOMY.

Fig. 1. Diagram of an open flower of Trillium showing the number and relative position of the parts : *s* sepals, *p* petals, *st* stamens in two whorls, *c* carpels each bearing two ovules.—Drawn with pen.

Fig. 2. Diagrammatic drawing of Marchantia to show the mode of branching, somewhat enlarged. As one branch of each new dichotomy soon distances the other, it produces the appearance of a main axis with right and left branches : *an* the extension into an antheridial branch, *ar* extension into an archegonial branch, *y* recent dichotomy, *o o'* older dichotomy in which *o'* is already perceptibly longer, *c* cupules which arise at the growing end of the midrib and are left upon its upper surface as the stem advances.— Drawn with pencil.

Fig. 3. Flower of radish, greatly enlarged and modified by the growth of Cystopus within it, natural size. The change induced by Cystopus is variable, sometimes single flowers are enlarged, as in this case, sometimes the whole cluster of flowers is changed when the individual flowers remain smaller. This example is larger than the average size.—Drawn with pencil.

Fig. 4. A small fruiting plant of Atrichum, × 2 : the stem bears scale leaves below and foliage leaves above, the base is clothed with rhizoids that simulate roots, *st* seta, *sp* capsule surmounted by the closely fitting calyptra. The distance the beak extends into the calyptra is indicated.—Outline drawing with pen.

Fig. 5. Flowering head from a vigorous male plant of Atrichum, × 2 : the difference between the perichætial and foliage leaves is well shown.—Drawn with pen.

Fig. 6. Pod (seed vessel) of Capsella, × 2.—Drawn with pen.

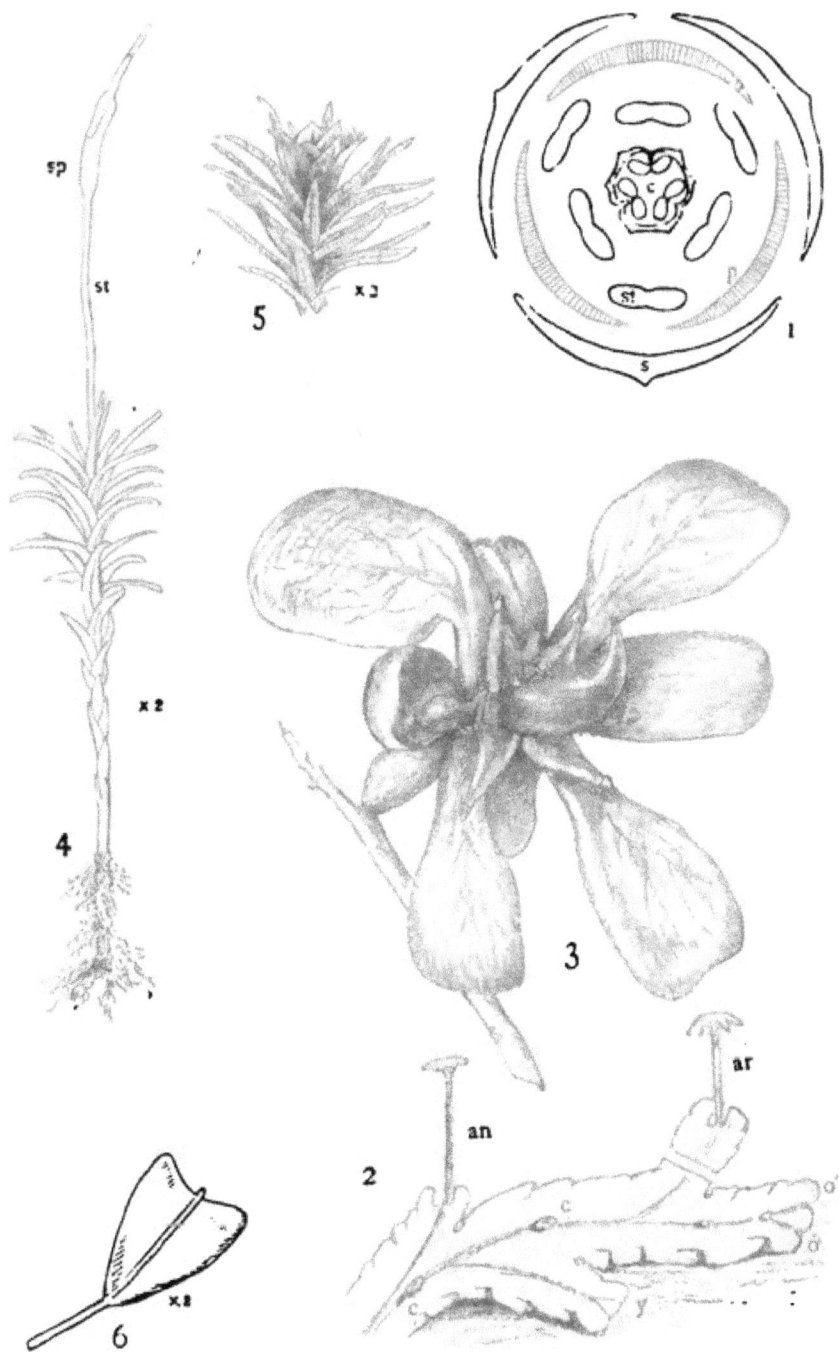

PLATE I.—Gross Anatomy.

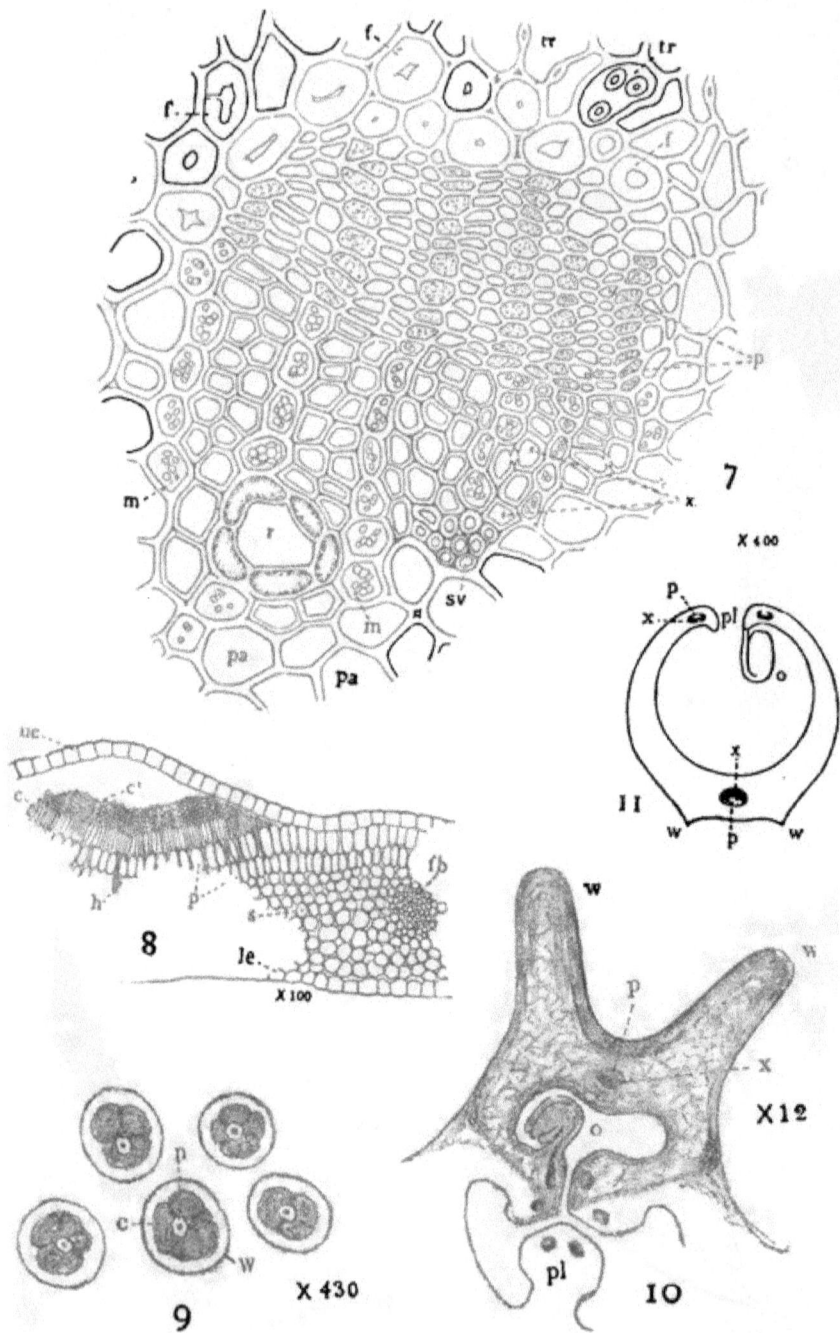

PLATE II. —Minute Anatomy.

EXPLANATION OF PLATE II.

ILLUSTRATIONS IN MINUTE ANATOMY.

Fig. 7. One of the pair of fibro-vascular bundles in a **leaf of** Pinus, × 400: *p* phloem, *x* xylem, *sv* group of spiral vessels (in other bundles they are often more scattered), *m m* rows of parenchyma cells forming **medullary rays** containing starch in the xylem and protoplasmic substances in the phloem, *r* **resin** duct, *f f* fibrous tissue with thick walls, small cavities and prominent middle lamellæ, *f'* fibrous cell with lateral pit, *pa* thin-walled parenchyma, *tr tr'* parenchymatous tracheides with bordered pits, *tr'* face view of the pits on an end wall. The other bundle of the same leaf was at the left side of this one.—Drawn **with pen.**

Fig. 8. Diagrammatic drawing **of a vertical section of leaf of** Capsella showing a sorus of Cystopus, × 100: *ue* **upper epidermis,** *le* lower epidermis, *p* **palisade parenchyma,** *s* **spongy parenchyma,** *fb* small fibro-vascular bundle, *h* hyphæ passing between **the pali**sade cells and terminating in *c* the conidiophores **which bear the** chains of conidia *c'*. **The epidermis is raised, but not yet ruptured,** above the sorus.—Drawn **with** pen.

Fig. 9. Cells of Protococcus after treatment with chlor-iodide of zinc, × 430: *w* the thick cell wall, *c* large chlorophyll bodies, *n* **nucleus with** central nucleolus.—Drawn with pencil.

Fig. 10. Diagrammatic drawing of a transverse section through the ovary of Trillium showing one entire carpel, which is shaded, and a portion **of the other two,** × 12: *w w* **the pair of** wings, *pl* the three placentæ meeting **in the** center **of the ovary,** *x* xylem and *p* phloem **of** the fibro-vascular bundles **of** which each carpel has one between the wings and **one in** each **placenta,** *o* ovule which receives a branch from **the** fibro-vascular **bundle of** the placenta to which it is attached.—Drawn **with pencil.**

Fig. 11. Diagram to illustrate **the** theoretical carpellary structure of Trillium, **representing a** single carpel **in** transverse section as in fig. **10, and with the same** lettering.—Drawn with pen.

INTRODUCTION.

I. INSTRUMENTS, ETC. ·

Following is a list of the instruments and appliances necessary and desirable for use with **this** manual. Those printed in italics **are** necessary ; **the** remainder are desirable **but** can be dispensed with.

GROSS ANATOMY.	MINUTE ANATOMY.
Hand lens,	*Compound microscope,*
Dissecting needles,	*Razor* **or** *scalpel,*
Razor or *scalpel,*	*Glass slips* (12),
Glass slips (3),	**Cover** *glasses* (24),
Cover glasses (6),	**Fine** *forceps,*
Drawing materials,	*Dissecting* **needles,**
Holder for lens,	*Drawing materials,*
Dissecting microscope,	*Blotting* or *filter paper,*
Fine forceps,	Camel's-hair brushes,
Fine scissors,	Fine scissors,
Camel's-hair **brush,**	Watch glasses,
Metric **rule.**	Dropping tube.

The *hand lens* should have **a** magnifying **power of** eight **to** fifteen diameters ; **one of ten or** twelve diameters **is** the best. Such a glass costs from 50 cents to $5.00, according to quality and mounting. One costing $1.00 will be found sufficiently good.

A *holder for the lens* may **be** constructed as follows and **answers every purpose of a** dissecting microscope :

Take a block of wood about 10 cm. long and 6 cm. wide. Fix upright in the middle of the block about 2 cm. from one end a bit of metal rod of 3 to 4 mm. diameter and 6 to 8 cm. high. Bore a hole a little to one side of the center of a smooth cork so that it will slide smoothly on this rod. Bore another hole at right angles to the first through which pass a wire of 7 to 8 cm. length. The free end of this wire may be bent into a loop or circle as may be desired to hold the lens.[1] The lens may be focused by sliding the cork up or down. Cheap loupe holders are also to be had of dealers in optical goods.

The mounted *needles* can be better made than bought. Take two number 8 " sharps," break off about one-third of the needle from the blunt end and grasping the remainder firmly with a pair of pliers, push the blunt end into a pine pen-holder or any suitable piece of soft wood till firm. The points of the needles should be kept sharp.

The *razor* should be of the best quality of steel without any stamped lettering or even etching on the blade, which should be at least 2 cm. wide. The best shape for the blade is to be ground flat on the under side (when held in the right hand with the edge toward one) and hollow on the upper. Next to this shape the "hollow ground" razor is best, provided the thin part of the blade is at least 12 mm. wide and not so thin as to be easily bent. " Extra hollow ground " razors have the blade too thin.

Glass slips with ground edges may be purchased of any dealer in microscopical supplies or they may be cut

---

[1] Modified from Kingsley, The Naturalist's Assistant, p. 83.

from clear window glass, or better from **photo-**
graphic plate; 76 mm. (3 in.) by 25 mm. (1 in.) is **the**
standard size.

Cover glasses **must be** bought. They should be 15 to
20 mm. in diameter or square. **No. 2** thickness is pref-
erable.

The *compound microscope* should be of good work-·
manship, which can be best secured by buying of some
reputable maker. A small low stand is to be preferred. It should have a good **fine** adjustment and be
furnished with two good objectives, viz., a 1 in., $\frac{3}{4}$ or $\frac{2}{3}$,
and a $\frac{1}{4}$, $\frac{1}{5}$, or $\frac{1}{6}$, and two eye-pieces, viz., A and **C**, or if
only one, a B. A combination of either eye-piece with
the 1 in., $\frac{3}{4}$, or $\frac{2}{3}$ is in this manual designated as a "low
power"; similarly, a combination with the $\frac{1}{4}$, $\frac{1}{5}$, or $\frac{1}{6}$
is known as a "high power." There should **also be a**
camera lucida, and a micrometer ruled in **fractions of a**
millimeter.

Fine forceps **should be of steel,** have very slender
bent points, and come together accurately. Those used
by dentists are excellent.

A large *camel's-hair brush* is desirable for dusting off
lenses. A small one with long hairs, which tapers to a
sharp point when wet, is very convenient for removing
sections from the razor. It should **be** mounted on **the**
small end of a pen-holder, in the large end of which **is**
a short needle. By sticking this in the table the brush
may be kept out of **the dust and always** handy.

Watch-glasses should have a flat **bottom to** prevent
tipping **too** easily. Plain individual salt-cellars answer
the purpose admirably.

A *dropping-tube* is a piece of small glass tubing drawn

to a point, with a rubber bulb on the larger end. They may be purchased in drug stores under the name of "medicine droppers."

Fine scissors may be either those made for anatomical purposes or small embroidery scissors. The latter answer most purposes well.

A *metric rule* is highly desirable. The student should have a pocket rule and should early familiarize himself with the metric system. Metric measures of various styles and prices may be obtained of the American Metric Bureau, Boston, Mass.

The *drawing materials* required consist of slips or a blank book[2] of unruled paper, hard and soft pencils, pens and ink. For ink drawings the paper may be either sized or unsized, rough or smooth, so long as the ink does not spread, but for pencil drawings the surface must be minutely roughened, and without sizing, in order that the plumbago may adhere well and give a soft effect. A quite hard pencil, No. 5, VH or HHHH, of artists' grades, is needed for tracing under the camera lucida, and one slightly softer than used for ordinary writing, No. 2, SM, or B, for completing drawings, especially those in gross anatomy. Ordinary steel pens, preferably those with slender points, and common black ink will suffice, but finer work may be done with lithographic pens and India ink.

II. REAGENTS.

The following reagents are necessary for the study of minute anatomy with this manual :

---

[2] If a book is used it must be so bound that it will lie flat on the table when open. The slips are usually preferred.

Alcohol,

Potassic hydrate,

Iodine,

Chlor-iodide of zinc.

Magenta,

Glycerine,

Sulphuric acid,

Potassic chlorate **solution.**

The *alcohol* used is the commercial article, 95 per cent. pure.

The *potassic hydrate* is a 5 per cent. solution of potassic hydrate in distilled water. Sodic hydrate will answer the same purpose. The "liquor potassæ" of the U. S. Dispensatory is of this strength and may be purchased of any druggist.

The *iodine* is prepared as follows: Dissolve 3 gm. of iodide of potassium in 350 cc. of distilled water; add 1 gm. of sublimed iodine. A weaker solution will be useful, viz., potassic iodide 3 gm., distilled water 500 cc., iodine 1 gm. The tincture of iodine diluted till it is a sherry brown color will answer in some cases, but is not so generally useful as the solution recommended.

Chlor-iodide of zinc may be prepared as follows: Dissolve metallic zinc in concentrated hydrochloric acid until the action ceases; evaporate to the consistency of syrup in contact with metallic zinc; saturate this with potassic iodide; add as much iodine as it will take up, with some excess.[3] It is better to keep the solution in a dark place, although in the majority of instances the proper reaction will be secured without this precaution.

Magenta is a solution of the aniline color of that name. It may be purchased of dealers in microscopical supplies or made as follows: Powder 1 gm. crystal-

[3] Poulsen and Trelease, Bot. Micro-Chemistry, p. 8.

lized magenta. Dissolve in 160 cc. distilled water, to which 1 cc. of alcohol has been added.[4]

The best commercial *glycerine* should be used. See that it is colorless and free from sediment.

A 75 per cent. solution of *sulphuric acid* should be prepared by mixing three volumes of c. p. sulphuric acid with one volume of distilled water, being very careful to pour the acid slowly into the water while stirring it.

The *potassic chlorate solution* may be prepared as follows: Dissolve 2 gm. potassic chlorate in 5 cc. nitric acid.

III. USE OF THE MICROSCOPE AND LENS.

The prime requisite in the use of any optical instrument is cleanliness: dirty lenses frequently defeat the very object of their use, namely, clearer vision. Before beginning to work with either the simple or compound microscope, see that the lenses are perfectly clean. When a lens needs cleaning, take a camel's-hair brush and brush away all particles of dust. Then wipe gently with a piece of soft unstarched linen or cotton—an old handkerchief is the best—breathing upon the surface slightly if necessary to remove the dirt. Too great care can not be taken to avoid scratching the polished surface of the lens; hence the least possible effective pressure should be used when wiping it. If properly handled after they have once been cleaned, lenses will seldom need any thing but brushing. One should avoid with the greatest care touching the surface of a lens with

---

[4] Huxley and Martin, Biology, p. 269.

the fingers, as finger marks are difficult **to remove : no** matter how clean the skin, the oil from it will adhere to the glass and can only be perfectly removed by wiping with linen moistened **with alcohol.**

When the **lens is held** in the hand **to** examine **ob-**jects, rest the hand holding the lens on the hand holding the object. They will then tremble together. The eye should be as close to the lens as possible in order **to** obtain a wider field of view.

In using the compound microscope the front only of the objective and both surfaces of both lenses of the eye-piece need cleaning. If the eye-piece be dirty there will be **specks in the** field of **view** when **there is** no object **on the** stage. These can **be** made more apparent **by turning** the eye-piece in **the tube while** looking through it. In like **manner** by partly unscrewing the eye lens and turning **it, it** may be discovered whether the **eye** lens **or** field **lens is** dirty. **If the** front of the objective be dirty **it will** be manifested by a dimness and want **of** definition **of** the outlines of objects, affecting **the** whole field of view.

In focusing with **the** high power of the compound microscope, first rack the objective down as close to the cover-glass as possible while watching it from one side. Then look through **the** tube, rack slowly back and **watch for** the coming **of** the object into view.

Never rack downwards while looking through **the** tube unless the object **be** in view.

Do not use the fine adjustment until the object **is** nearly in focus with the coarse.

Raise **the objective** slightly before placing or removing **a slide.**

An object is examined by "direct" light when it is examined by the light which falls upon its surface without passing through it. This is the common method with the hand lens.

An object is examined by "transmitted" light when the light passes through it before entering the eye. This is the common method with the compound microscope. Ordinarily, when transmitted light is used, direct light should be cut off as far as possible.

An object is examined by "oblique" light when the light passes through it so obliquely that only that refracted by the object enters the eye. It therefore appears light against a dark ground.

IV. SECTION CUTTING.

Sections.—A section is a very thin slice taken from the interior of any organ. It should be of as nearly equal thickness in all parts as possible. The term "slice" is used to designate a thin piece cut from the surface of any organ.

By a transverse section is meant one at right angles to the long axis of the object. Unless care is exercised the surface from which the sections are being cut will become inclined. Especially is this likely when the object is large or is supported in pith. The pith stick should be trimmed down at the end so as to leave only enough to support the object. The chief cause of the tendency to become inclined is that the under side of the razor is not flat; hence the larger the object, the more likely the transverse sections are not to be truly transverse.

By a longitudinal section is meant one which is

parallel to or includes the long axis of the object. It is evident that longitudinal sections of all cylindrical objects may be either radial or tangential. A radial section is one lying in the plane of a radius. A tangential section is one parallel to a plane tangent to the cylinder.

Longitudinal sections are much more difficult to make than transverse and they are nearly or quite useless unless truly longitudinal.

The razor.—The secret of making good sections lies in having and keeping a sharp razor. No amount of skill can make a dull razor cut a thin section.

The edge of the razor must be free from nicks. This can be determined by looking at the profile of the edge against a bright light with a lens. Nicks, if small, can easily be taken out on a hone.

The razor should be stropped often. It is easier to keep it from getting dull than to sharpen it after it has become so. If its edge is free from nicks and it will cut a hair of the head 2 cm. from where it is grasped by the fingers, it is in good condition.

After using the razor be careful to see that no moisture or plant juices are left on the blade; they will surely rust it if allowed to remain.

Holding specimens.—Large specimens of which sections are to be cut may easily be held in the fingers. They should be held vertical, grasped by the fore-finger and thumb of the left hand so that the razor blade may rest on the corner of the fore-finger, and the remainder of the hand be out of the way below.

Small objects should be placed in a piece of elder or sunflower pith in which a median longitudinal slit has been made, deep enough to allow the ends to spring as far apart as necessary to receive the specimen, between which it is to be firmly held. The pith is then to be grasped as a large specimen for cutting. If alcoholic specimens are being used the pith should be previously soaked in alcohol, and if fresh material, in water.

Cutting.—Grasp the razor firmly with the right hand where the blade joins the handle, bracing the blade by resting the thumb against the tang. Hold the razor horizontal, rest the under side of the blade against the corner of the fore-finger and cut toward you, pushing the razor from point to tang or drawing it in the opposite direction, using as much of the blade in cutting as possible.

If the object be flat and thin, as a leaf, let the razor edge pass through it at an angle of 20° to 30° to its length.

If alcoholic specimens are being cut, the razor blade should be flowing with alcohol. The oil usually on the blade from stropping will prevent the alcohol from running off, unless the blade be considerably inclined. If fresh material is used the razor should be dipped in water. The object of the operation in both cases is to prevent the section from becoming dry. Should it do so, it will inevitably contain air bubbles when mounted, which will unfit it for examination.

Removing the sections.—The most convenient thing for removing sections from the razor is a small camel's-hair brush, which, when wet, tapers to a sharp point.

With such a brush a section may easily be picked up from the water or alcohol, in which it ought to be floating on the razor, and transferred either to the slide or to a watch glass. An easier way of removing sections which are to be transferred to a watch glass, is to wash them down to the point of the razor, and then dip the point of the blade in the liquid in the watch glass.

Always cut a number of sections—half a dozen or more—at once. One or more may prove good.

V. MOUNTING.

Previous to mounting any specimens, it must be seen that the slide and cover glass are perfectly clean. Nothing is better for cleaning slides and covers than a clean linen handkerchief, which should be used for this purpose alone. The cleaning of the slide is a simple operation; the cleaning of the cover requires more care, to prevent breaking. Having dipped the cover in clean water, take it between the thumb and fore-finger, over which a single thickness of the handkerchief has been thrown. Wipe gently, using the fingers of the other hand to keep it in place. The surfaces of the cover should be perfectly cleaned, so that when light is reflected from them, no oiliness or dust is visible. Having cleaned the cover, lay it down in some clean place, with one edge projecting slightly, so that it can be readily picked up, or stand it on edge against some support.[5] Having placed the desired specimen

[5] A very convenient receptacle for covers, whether clean or dirty, is made by sawing several grooves in a block of wood, and nailing across the ends of the grooves a thin strip. In these grooves the covers rest on edge. A similar arrangement is useful for slides.

in the center of the slide in a drop of water, grasp the edge of the cover firmly with the fine forceps, breathe on the under side, hold it in a slanting position over the drop of water, place the lower edge in the edge of the drop, and lower it gradually on the water. The condensed moisture of the breath insures more ready contact of the water with the cover, and lowering the cover slantwise gives opportunity for the escape of air from under it.

If air bubbles appear in the mounting, they are due to one of two reasons: either (1) the cover glass was not clean, or (2) it was dropped instead of being lowered to the slide. Of these the first is the more common cause of air bubbles. They may sometimes be removed by lifting one edge of the cover with a needle, while the other is prevented from slipping, and then lowering again. Sometimes it will be necessary to remove the cover, clean and replace it.

If the bubbles appear in the specimen itself, they are probably caused by allowing the section to dry partly before mounting. They may usually be removed by taking off the cover and treating the specimen with alcohol.

The worker should not be content to let bubbles remain.

Another difficulty is sometimes encountered, when it is attempted to mount several sections under the same cover, in the floating out of one or more. This is usually due to an excess of water. The remedy is to take up the cover, absorb some of the water with filter paper, and re-cover.

After covering specimens, soak up the superfluous

liquid sufficiently to prevent the cover floating when the slide is inclined.

VI. APPLYING REAGENTS.

Stains are most conveniently applied by placing a drop of the liquid at the edge of the cover, and allowing it to run under, hastening the process when desired by placing a strip of filter paper at the opposite edge. If the stain does not reach all parts of the specimen, the cover glass may be slightly raised. It is quite important in many cases to watch the action of reagents. In such a case they should be applied with the stage of the microscope horizontal. Time may be saved when it is necessary to examine specimens in potash, by placing a drop of potash on the slide, and mounting directly in that medium. In all cases, as soon as the specimens have become clear, the potash should be washed out with water: otherwise the cell walls swell excessively, and many points become indistinguishable. It should be remembered in examining specimens treated with potash that many cell walls are somewhat swollen, and that the longer they remain the greater the swelling becomes.

Glycerine is one of the most useful media for clearing, and at the same time preserving specimens for prolonged examination. Whenever it becomes desirable to preserve specimens from one day to another, a drop of glycerine should be applied to the slide, so that it just touches the edge of the cover, and the slide laid away in a horizontal position. As the water or alcohol evaporates, the glycerine will run under the cover. The excess may be wiped off with a damp rag

after a few hours. Specimens may be mounted directly
from alcohol or water in glycerine, but the saving in
time will not be material, unless it is known that the
specimens are good, before mounting. Care must be
taken that the glycerine does not overspread the cover,
which, under such condition, must be taken off and
cleaned. Most specimens may remain in glycerine any
length of time without deterioration, and will become
clearer and clearer all the time. Care must be taken to
keep the specimens thus preserved free from dust.
They must be handled cautiously, lest the cover be
shoved off. If desired, specimens which have been
preserved in glycerine, may be permanently mounted,
by simply running a ring of shellac cement around the
cover.[6]

The greatest care must always be exercised to prevent reagents from coming in contact with the stand
of the microscope or the lenses, as most of them attack
the lacquer of the brass work, and some the brass.
The chief danger arises from a failure to remove the
excess of the reagent, which then collects at the lower
edge of the slide when the microscope is used in an
inclined position, and runs off on the stage. Sulphuric
acid behaves in the same way even when the excess
is once removed, if it is allowed to remain any length
of time, because it absorbs moisture from the air.

Boiling specimens in the potassic chlorate solution
should not be done in the same room with the microscope as the liquid and its fumes are intensely corro-
sive.

---

[6] For directions for making this cement, see *Am. Mo. Mic. Jour.*, v.
(1884), p.131. Similar cements may be bought.

VII. CARE AND USE OF MATERIAL.

Throughout the directions for laboratory work it is understood that material preserved in alcohol will answer unless otherwise stated. In many cases only alcoholic specimens are usable and in other cases only fresh specimens.

Do not tear up specimens needlessly.

Examine a specimen thoroughly and see as much as possible before dissecting.

Do not begin dissecting a part until it is decided what to look for and where to look for it.

Be economical ; chiefly because it is a good habit, secondarily because material costs time, or money, or both.

Save the pieces ; they may be useful in future work : it is easy to throw away ; it is more difficult to gather.

Preserve all sections and other preparations until the study of the plant is completed.

When the specimens are mounted in water be careful lest they become dry by the evaporation of the water. It can be most conveniently replaced by placing a brush charged with water at the edge of the cover opposite the area of air. As soon as the air is displaced the brush should be removed.

When studying particular tissues in a section the thinnest parts of each tissue should be selected. It is rare that a section is so uniform that the tissues are equally well shown in all parts of it, and different tissues must not infrequently be looked for in different sections. It is best therefore to look well over the specimens before settling to the study of any tissue.

In order to obtain a clear conception of the shapes

of the cells of a particular tissue, it is indispensable that
the student carefully compare the transverse and longitudinal sections of the cells. Moreover the longitudinal sections must be compared with the transverse to
determine their position.

It frequently becomes necessary to examine a tolerably thick object. In such a case, very different views
of the object will be obtained as the focusing screw of
the fine adjustment is moved. It must be remembered
that a good objective gives a clear image of only a
single plane at one time, though adjacent images
modify this somewhat. Hence it is easy to determine,
knowing in which direction the objective is moved by
the focusing screw, whether one object is above or
below another.

The use of the fine adjustment must be learned as
soon as possible and must be assiduously practiced.
The finger should be kept on the fine adjustment most
of the time when using high powers, and nothing
allowed to escape the vision which the fraction of a
turn would reveal.

VIII. DRAWING.

In the systematic examination of an object two kinds
of memoranda should be made, descriptions and
drawings. The value of the former is usually conceded,
but that of the latter is often deemed too slight to re-
pay the trouble. The importance of drawing can not,
however, be too strenuously urged, and the difficulty
and tediousness of execution, which will largely dis-
appear with practice, should never be offered as an
excuse for its neglect.

Drawing may represent the object **with** various degrees of fidelity. **At** one extreme is the diagram (see fig. 1), which only aims to give the relative positions or sizes **of** the several parts, or some other feature. **At** the other extreme the drawing is **as** close a **counterpart** of the object seen as the person who draws **it is** capable of producing (see fig. 3). Whether a particular object shall be drawn in one way or the other, **or** in some intermediate way, must be determined by the nature of the object and the end to be attained by the study.

The usual tendency is to make drawings too small; they should be large enough to show all parts distinctly without close scrutiny.

Drawings may usually be satisfactorily made in outline, or with very little shading, as in fig. **4 or 6.** They are **most** easily drawn with a soft pencil **on** heavy, unsized and slightly calendered paper, producing the effect in fig. 3 or 10, but are not permanent; rubbing readily defaces them, unless treated to a fine spray of colorless shellac dissolved in alcohol, which may be applied with an atomizer, such as is used for perfumery. Ink drawings are to be preferred for their durability and distinctness. When **ink** is used, the main features of the drawing should first be lightly sketched with a hard pencil, and the pencil marks erased after the ink is dry.

Drawings in gross anatomy should be the exact size **of** the object, or **some** multiple of **it.** Record the amount of linear enlargement by **a** number placed **at one** side of the drawing with an oblique cross prefixed.

In **the directions for laboratory** work in gross

anatomy the number of drawings has been mostly left
to the discretion of the student.

In minute anatomy the points at which drawings
may most profitably be made are carefully noted. In
many instances, however, it is so difficult to secure a
wholly satisfactory section to show certain structures,
that they should be drawn whenever found in good
condition, without regard to the directions.

Drawings in minute anatomy may be either free-
hand or with camera lucida. In free-hand drawing the
student is especially cautioned against making them
too small, which is a very common fault. In the outlines for work it is expected that accurate drawings be
made unless a diagram or diagrammatic drawing is
called for. A diagram (fig. 11) shows only a single
special feature, or at most two or three, while a diagrammatic drawing (fig. 8 or 10) shows all the chief
features, but does not take note of smaller matters,
such (e. g.) as distinguish the several cells of the same
tissue. When an accurate drawing is to be made, each
individual cell should be drawn as carefully as if it
were the whole object. When an accurate drawing
includes considerable tissue, time may be saved by
indicating the boundaries between the tissues by dotted outlines, and only cells enough filled in to show
the character of the tissues.

In order to draw to scale with the microscope it is
necessary to use a camera lucida. The magnification
is thus determined : place a micrometer on the stage
of the instrument in the same position as an object,
adjust the instrument as for drawing, and laying a
common rule on the drawing paper read off the dis-

tance that the image of one division of the stage micrometer covers on the rule. If, for instance, a tenth of a millimeter of the stage micrometer covers five centimeters (five hundred tenths of a millimeter), any drawing under the same adjustment will be magnified five hundred times. Always mark the number of times magnified at the side of the drawing as in gross anatomy, thus, × 500.

The distance from the drawing-paper to the reflecting surface of the camera lucida should be about the same as from the latter to the outer lens of the object glass, in order that the drawing may properly represent the magnifying power of the instrument. **Ten inches has been adopted as the standard length of tube.**

Trace the image first with a hard pencil, and then go over it with ink before the object is removed from the instrument in order to correct any errors made by the pencil.

It is not an easy matter to draw accurately with the camera lucida, owing to the difficulty in seeing both the image and the pencil point distinctly at the same time. Much depends on the relative amount of light received by the eye from the instrument and from the drawing-paper. If the pencil point does not show clearly, there should be more light on the paper, and if the image is not clear, more on the object.

Invariably accompany each drawing with a full explanation.

IX. BOOKS OF REFERENCE.

It should be the aim of the student to find out all that he can about the plant in hand with as little assistance as possible or without any.   This requires patient and thorough work.   When done, however, and drawings and notes have been fully recorded, it will be advantageous to compare the work with the published observations of others, and if any points have been overlooked or misunderstood, to go over the ground again.

The following general treatises will be found suitable for preliminary consultation, and when possible should be constantly at hand on the laboratory shelves : Gray's Structural Botany, Goodale's Physiological Botany, Bessey's Botany for High Schools and Colleges, Sachs' Text Book, 2nd Eng. edition, Prantl and Vines' Text Book of Botany, DeBary's Comparative Anatomy of Phanerogams and Ferns, Strasburger's Das botanische Practicum, Poulsen and Trelease's Botanical Micro-Chemistry.

If the student becomes interested in any particular direction, the references given in the annotations, together with those to be found in such of the works just named as may be at hand, will usually give him a fair start in tracing the literature of the subject, and becoming acquainted with what has already been ascertained in regard to it. This will indicate wherein present information is defective, and enable him to direct his labors toward a profitable increase of the total sum of knowledge.

The references have been preferably to works most likely to be at the student's command, whenever these have contained a sufficiently full treatment, this doubtless tending more to accomplish the desired object of interesting the student and leading him on to independent work, than references in all cases to the original sources of information. Less accessible works have often been cited to introduce the student at once to the most complete treatment of the subject. A few citations are for the sake of authority.

Many of the memoirs and articles cited in apparently inaccessible foreign journals and proceedings of societies may, however, be bought separately of foreign dealers (R. Friedländer & Sohn, Berlin N. W., Germany, and many others). A very moderate outlay will thus enable one to consult numerous valuable writings.

No apology need be offered for referring in an elementary work to writings in foreign languages, for unless the student carries his researches outside this manual he will have no occasion to use them, and if he does do so he can not go far without being obliged to use them. It is not often possible in fact to treat a subject exhaustively in the departments of botany covered by this handbook without a knowledge of German and French writings at least.

But if the references given among the annotations are never used, they will still serve a good purpose in impressing upon the learner that he is only upon the threshold of the study, and that the facts which he seems to be gathering so thoroughly are in most cases to be found more fully and accurately set forth in the great storehouse of learning beyond.

GREEN SLIME.

Protococcus viridis Ag.

PRELIMINARY.

THE plant selected to illustrate the simplest phase of vegetable life is found in all parts of the United States, and even throughout the world. It grows upon the surface of various objects, being often so abundant as to give them a conspicuous green color, especially upon the north side of old fences, barns, and the trunks of trees, becoming more noticeable after a few days of damp weather. There are several other closely related species that may be used, in fact almost any unicellular green plant will answer, but this is the one most likely to be gathered. Some kinds of unicellular plants, like *Glæocapsa*, have a sheath or envelope outside the cell proper, not found in Protococcus, a fact to be borne in mind by the student if such plants are used. Pieces of bark or wood bearing the alga may be kept dry for use, and will give a fresh appearance when moistened with water, and even retain vitality for a year or two.

It is quite likely that the plants known under the name of Protococcus are but early forms of some more complex algæ[1], but, however this may be, they serve

---

[1] Bessey, Botany. p. 219; Wood, Fresh-Water Algæ of North America, p. 10; Sachs, Text-Book of Botany, 2nd Eng. ed., p. 248; Cienkowski, Bot. Zeit. 1876, p. 17.

quite as well as any to illustrate the simplest kind of plant life.

To complete the following study it will be necessary to have pieces of wood bearing the Protococcus; iodine; chlor-iodide of zinc; and alcohol.

LABORATORY WORK.

GROSS ANATOMY.

Taking either a fresh or dried specimen, notice
1. The *color*.
2. The *evenness* with which the plant overspreads the supporting surface.

Using a lens, notice
3. The pulverulent *appearance*, as if dusted or sanded upon the surface.
4. The appreciable thickness reached in some spots, causing it to separate in scales in the dried specimen.

Mount, and observe
5. The *dust-like particles*[1] into which it separates.
6. The varying size of the particles.

Place a piece of bark with the Protococcus in a small quantity of alcohol, after an hour or more notice
7. The color imparted to the alcohol by the coloring matter of the plant, the chlorophyll.[2]

MINUTE ANATOMY.

Under high power, notice

---

[1] Care must be taken not to confound them with air bubbles, which are often numerous when a dried specimen is used.

[2] Some less common forms of unicellular algæ are red or purple from additional coloring matter.

1. The individual cells; either single or associated in families.
2. The *size* of the cells ; some small, some several times larger.
3. The *shape ;* when free and when in families.
4. The cell *contents ;* more or less granular, and always green from the presence of chlorophyll.
5. The colorless cell-wall surrounding each cell.

Press upon the cover-glass with a back and forth movement, and the walls to many of the cells and cell-families will be ruptured and their contents ejected, when the wall can be easily studied.

Stain with iodine and notice

6. The brownish-yellow color given the contents of the cell, showing the presence of protoplasm.

Stain a freshly mounted specimen with dilute chloriodide of zinc, and after an hour or two[4] notice

7. The two to several closely packed bodies of definite outline, usually overlapping, forming the green part of the cell, the chlorophyll bodies, best seen in the largest, single, round cells.[5]

8. The small round body nearly in the center of the cell, or in recently divided cells near the partition wall, the nucleus.[6]

9. Occasionally a clear space between the chlorophyll bodies and the cell wall, occupied by the *protoplasm.*

10. Draw a few cells showing chlorophyll bodies and nuclei.

---

[4] If the cells are properly stained they will usually remain green, but of a brighter and more bluish hue.

[5] There is danger of mistaking delicate partition walls of young cells. which the reagent has thickened and made visible, for the boundaries of the chlorophyll bodies.

[6] Under higher power yet a central dot to the nucleus, the *nucleolus,* may be detected.

11. The *cell multiplication :* examine various specimens and trace the successive stages in the division of a single cell to form a cell family.

12. Illustrate the cell multiplication by drawings.

ANNOTATIONS.

Protococcus is a unicellular plant, for each cell performs individually the various functions pertaining to plant life ; and this is true whether the cells remain single or become associated into small families.

The cell is the unit from which all plants, however complex, are built up.

The most essential part of the cell is the protoplasm, a colorless semi-fluid substance, which in this instance is masked by the green chlorophyll. It is the only really living, active agent in this, as well as in all other plants. Its presence here is made manifest by the characteristic yellowish-brown color given by iodine.

The nucleus (see fig. 9 *n*) is a special form of the protoplasm to be seen in most plant-cells. As its division usually precedes that of the cell, it has generally been regarded as in some way necessary to the latter process. The investigations of Schmitz, Strasburger[7] and others go to show, however, that the two processes are distinct, and that the nucleus, instead of being related to cell division, holds an intimate and probably essential relation to the life of the protoplasm.

The protoplasm takes on another form in the chlorophyll bodies[8] (see fig. 9 *c*). These consist of a proto-

---

[7] Zellbildung und Zelltheilung, p. 371.

[8] Cf. Strasburger, Das botanische Practicum, p. 350 ; Schmitz, Chromatophoren der Algen.

plasmic body containing the green chlorophyll pigment. The surrounding protoplasm by the aid of the chlorophyll is able to convert inorganic into organized matter, a function wanting in all animals, with the exception of a few of the lowest, like Hydra and Euglena, and also wanting in some plants, e. g. fungi and colorless parasites.

The solid, firm, and nearly colorless cell-wall is a product of the protoplasm consisting essentially of cellulose, and serves as a protection to the protoplasm. The fine granules seen in the protoplasm, are largely food materials produced by the cell in excess of what the present needs require.

The multiplication of the plant by cell-division is a very common method throughout the vegetable kingdom.[9] The nucleus first disappears and two nuclei are formed in its stead. The protoplasm then divides itself, keeping a nucleus in each part, and a wall is formed between. The two cells thus produced soon attain the size of the original cell, when they in turn divide into two, but usually by a partition at right angles to the last, and so on. The cells thus formed either soon become separated, or retain a mechanical union.

Another method of multiplication is by the production of zoospores.[10] The plastic contents of a cell, either as a whole or divided into several parts, escapes from the cell wall, each mass pushes out a pair of delicate protoplasmic filaments or cilia, which moving

---

[9] Cf. Bessey, Botany, p. 36, for a statement of the different methods by which new cells are formed.

[10] Cf. Huxley and Martin, Elementary Biology, p. 12, 15 ; Howes, Atlas of Elementary Biology, p. 74, pl. xviii.

rapidly back and forth propel the naked protoplasm through the water. The motion and form give a strong resemblance to some of the simplest animals, hence the name of animal-like spores. After a time they come to rest, draw in the cilia, secrete a cell-wall, and become ordinary Protococcus cells. Sometimes the protoplasm does not free itself from the cell wall, but contracts somewhat, the cilia are protruded through the wall and the mass propelled as just stated. The production of zoospores at a specified time, as for a class demonstration, is attended with so much uncertainty that their study has been omitted from the laboratory work. This method of asexual multiplication will be studied later under more favorable conditions in Cystopus.[11]

---

[11] At p. 47.

DARK GREEN SCUM.

Oscillaria tenuis **Ag.**

PRELIMINARY.

THE color of Oscillaria, almost any species of which may be used, is generally sufficient to enable one to distinguish it at sight. Its dark blue-green is in marked contrast with the yellow-green of most other plants which form scums. It is very common on stagnant water, often forming patches of scum thirty centimeters (a foot) or more in diameter, which becoming loaded with dust finally sink to the bottom. It is also very common about watering troughs, along street gutters, at the outlet of drains, on wet rocks, giving them a slippery surface, in the greenhouse, and especially in water containing a small amount of garbage. It can usually be grown indefinitely in an open jar, by supplying the water as it evaporates, or with less trouble, when once established, in an unstoppered bottle, in which a small twig or flower stem of some sort is inserted to provide nutriment. The plants are often to be found in winter in as good condition as in summer. The study should be made upon growing plants when possible, but specimens dried on paper or mica will serve quite as well, except to show the oscillating movements, which are characteristic of the group to which Oscillaria belongs.

Only the following **material** is necessary **for the** study: fresh **plants, or** in their **absence dried** speci- mens; **a dried mass half** as **large as a pea; and** alco- hol.

LABORATORY WORK.

GROSS ANATOMY.

1. Examine a small mass of the living plant which has been allowed to remain undisturbed for several hours in a watch-glass of **water** ; notice
   a. The deep **blue-green** *color.*
   b. The **hair-like** unbranched *filaments,* radiating from the central mass.
2. **Sketch the plant as it** appears in the watch-glass.
3. **Mount a fragment and observe the uniform** *diameter* **and** *appearance* of the filaments.

Pulverize **a mass of the plant that has been thoroughly** dried, place **in a** test-tube **or vial** with **nearly** twice the bulk **of water, and** after **ten to** twenty-four hours notice

4. The color of the solution when seen by transmitted light and the **very** different color by reflected light, **indicating** the presence of **phycocyanine.**

Pour off the supernatant water, add **the same amount of** alcohol instead, and after an hour **or more** notice

5. The yellow-green **color** imparted **by the** *chloro- phyll.*

MINUTE ANATOMY.

A. GENERAL CHARACTERS. Under a low power, notice

1. The *color.*

2. The numerous *filaments* of uniform diameter, destitute of branches.

3. Study the *movements.*

B. THE INDIVIDUAL FILAMENT. Under high power, notice

1. The *structure* in detail, as follows :
   a. The rounded *extremities* of uninjured filaments.
   b. The outline of an uninjured *apex*, whether attenuated or not, and whether bent to one side or straight.
   c. The delicate lines of the *partition walls* crossing the filament and dividing it into very small cells.
   d. The comparative *length* and *breadth* of the cells.
   e. The granular *contents*, and their distribution in the cell.[1]
   f. The delicate colorless sheath to be seen extending beyond the green cells at some torn end of a filament, and on which may sometimes be detected transverse lines indicating the former position of the end walls of the cells.

2. The *turgidity* of the cells: notice that
   a. The transverse walls in an uninjured filament are ·plane, while·
   b. The last cell of an injured filament is bulged outward, making the outer transverse wall convex, the pressure from within not being counterbalanced from without.

3. Draw one or more filaments.

---

[1] In some species the granules are collected along the partition walls.

ANNOTATIONS.

If the structure of Oscillaria be carefully compared with that of Protococcus more points of resemblance will be found than appear at first sight. New cells are formed by the process of division, as in Protococcus, but the partition walls are always parallel and in one direction, which disposes the cell families in filaments. The individual cells have thin walls, the office of protection being relegated to the sheath. The sheath, which is formed from the outside walls of the cells by a modification of the outer portion, is a structure that is mostly confined to certain groups of the lower plants, although it has some analogies with the cuticle of the higher plants. The protoplasm is homogeneous, and not differentiated into chlorophyll bodies and nucleus as in Protococcus; chlorophyll is, however, present, evenly distributed through the protoplasm, but no nucleus has yet been discovered. The study of the protoplasm and chlorophyll is much obscured by the presence of the peculiar coloring matter, phycocyanine, characteristic of the *Cyanophyceæ* to which Oscillaria belongs. It is this that gives the deep blue-green color to the plants, enabling one to distinguish them at sight. It is insoluble in alcohol, but soluble in water when the plants are dead, while chlorophyll is soluble in alcohol, but not in water; hence, digesting the dead plants with water removes the phycocyanine, and digesting with alcohol removes the chlorophyll.[2] This blue color is often seen on the sides of vessels in which Oscillaria has remained so long as to die, and also staining the

---

[2] Cf. Sachs, Text-book of Botany, 2nd Eng. ed., p. 246, 766.

herbarium sheets on which specimens have been dried.

The cells are assisted in keeping together by the investing sheath, into which they are packed like a roll of lozenges in their case. This structure, together with the community of action exhibited in producing the peculiar oscillating and nutating movements, makes it evident that the cells of each filament have a certain dependence upon each other, although at the same time each is capable of independent existence. It may be that the smallness of the cells and the thinness of their walls is in some way correlated to this unity of function. It is not yet definitely known how the movements in Oscillaria are produced.[3]

Turgidity is a characteristic of living cells. It is the means by which the soft parts of plants are enabled to keep their form, and otherwise to serve their purpose. It is brought about by the strong imbibition power of the protoplasm, causing water to be taken up until a considerable internal pressure is created.[4]

---

[3] Engelmann has discussed several theories, and suggested that the movements are brought about by vibratile thread-like extensions of the protoplasm through the cell walls. Bot. Zeit. 1879, p. 49 According to Hansgirg it is due to an osmotic action of the protoplasm. Bot. Zeit. 1883, p. 831.

[4] Cf. Bessey, Botany, p. 166.

COMMON POND SCUM.

Spirogyra quinina **Kütz.**

PRELIMINARY.

THE members of this genus are abundant in stagnant water everywhere, forming bright yellow-green scums of great extent, sometimes diffused beneath the surface, or in running water attached to stones. They may be readily distinguished from all other scum-producing plants, except from a few of their close allies, in having a slippery feel, and being composed of long unbranched filaments, which string out like wet hair when withdrawn from the water. The allied kinds, which can not be separated by this simple test, will at once be distinguished when placed under the microscope by possessing no spiral chlorophyll bands as in Spirogyra. When growing vigorously the masses of Spirogyra are an intense light green ; when beginning to fruit they turn brown, and look very uninviting ; but as the characters which distinguish the species are largely drawn from the fruiting condition, the collector soon learns to regard these unsightly objects with favor.

The vegetative condition may be found at any time during the warmer portion of the year. The fruiting condition occurs from early spring to June and July, and sparingly during the remainder of the warm season.

The species usually grow intermixed, and almost any

gathering will answer for the present study, as *S. longata* Vauch., *S. majuscula* Kütz., and similar kinds have been kept in mind as well as *S. quinina* in drawing up the outline for laboratory work.

Spirogyra may be grown in the laboratory, and the vegetative condition kept always at hand, by using a rather deep vessel with opaque sides, and occasionally dropping in a small piece of peat which has been thoroughly boiled and afterward saturated with the following nutritive solution: 1,000 cc. of water, 1 gm. potassic nitrate, .5 gm. sodic chloride, .5 gm. calcic sulphate, .5 gm. magnesic sulphate, and .5 gm. finely pulverized calcic phosphate.[1] The last, for which burned bone may be used, is only sparingly soluble. If running water can be conducted through the jar containing Spirogyra, so that the water in it may be slowly changed, the peat and nutritive solution can be dispensed with. The fruiting plant may be preserved in fair condition for study in a fluid of equal parts of glycerine and alcohol.

The requisites for study are thrifty growing plants; fruiting plants, fresh if possible; alcohol; glycerine; and iodine.

LABORATORY WORK.

GROSS ANATOMY.

A. GENERAL CHARACTERS. Notice

1. The yellow-green *color* as seen in mass.

---

[1] Sachs, Vorlesungen über Pflanzen-Physiologie, p. 342.

2. The slippery *feel*, when the plant is taken between the fingers.

Float a small amount of material in water over a white surface, and observe

3. The fine unbranched *filaments* of which it is composed.

4. Their uniform *diameter*.

5. Their *length*.

Place some in alcohol, and after some time notice

6. The color imparted to the alcohol by the *chlorophyll*.

B. Mount a few filaments, and notice the single row of alternating light and dark dots, indicating the single row of *cells*. This can not be seen in all specimens.

C. THE FRUITING PLANT. Mount a few filaments from a fruiting mass, having them well separated on the slide, and search for

1. Paired conjugating filaments, some cells of which are empty, some with dark colored dots, the zygospores, and a few often remaining unchanged from the vegetative condition. [2]

MINUTE ANATOMY.

A. GENERAL CHARACTERS. Under low power, notice

1. The indefinite *length ;* if traced to the end, the filament will probably be found broken.

---

[2] The presence of small particles of dirt and other débris makes it difficult to distinguish the zygospores and conjugating filaments with certainty, and it is always best to verify the observation with the compound microscope, if possible.

2. The uniform *diameter.*

3. The cell *contents ;* colorless, except the conspicuous green chlorophyll bands.

B. THE INDIVIDUAL FILAMENT. Using a high power, notice

1. The *shape* of the cells.

2. Their relative *length* and *breadth.*

3. The *cell wall :*
    a. The *lateral walls ;* parallel and without markings of any sort.
    b. The *end walls ;* at right angles to the longitudinal axis, and plain (unless slightly nodulated or infolded, which occurs in a few species).

4. The absence of any visible *sheath*, although the presence of at least a thin one has been demonstrated by the slippery feel.

5. The *cell contents.*
    a. The *chlorophyll bands*, taking a spiral course from one end of the cell to the other, passing near the periphery. Note
        i. The number in each cell.[3]

---

[3] When a cell is crowded with chlorophyll, the following method may be used to advantage in determining the number of bands : count the number appearing to cross the band *ab*, between the point *a*, the upper profile view, and the point *b*, the lower profile view ; this number plus one will be the number required. The diagram shows a cell with four bands of chlorophyll.—From *Bot. Gazette*, ix., p. 13.

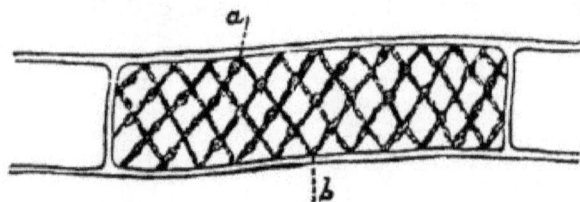

ii. The number of turns of the spiral.

iii. The surface, the crenulated and wrinkled margin, and the turned up edges of the band forming a more or less flattened **V** in optical section. To obtain a complete conception of these particulars, first focus upon the peripheral surface of the band, *i.e.*, upon the upper (outer) surface of the part nearest the eye, then focus upon the axial (inner) surface, and finally examine the profile of the band seen on the right or left of the cell.

iv. The *nodules* at varying distances along the median line of the band. Stain with iodine and note

    α. An outer ring which is more deeply colored, starch,[4] and

    β. A central light spot, pyrenoid. Both are best seen when but faintly colored.

v. The yellowish brown color finally imparted to the chlorophyll band.

b. The feeble brownish color given to the remainder of the contents of the cells, deeper along the periphery.

Run under glycerine on the same slide, and note

c. The contraction of the colored protoplasmic part, and its separation from the cell wall.

d. In unstained cells presenting the least obstruction from the chlorophyll bands, search for a colorless irregular body with radiating arms, near the center of the cell, the nucleus. This is difficult to demonstrate in some species, but easily seen in others.

---

[4] Unless the plants have been in sunlight the preceding part of the day the test for starch may not be fully successful.

e. The rounded, usually much brighter body imbedded in the nucleus, and occupying a considerable part of it, the nucleolus.

f. Draw one or more cells showing all parts noticed.

6. The *turgidity* of the cells, shown by the considerable convexity of the last end wall of a broken filament, which is repeated in lessening degree by the walls of successive cells until a point is reached where the pressure on opposite sides is equal, and the wall remains plane. Illustrate with a sketch.

C. THE FRUITING PLANT. Under low power, notice

1. The filaments lying side by side in pairs, held together by conjugating tubes.

2. The irregular outline of the filaments, caused by the uneven lateral expansion.

3. The varying character of the contents of the cells : some with distinct bands of chlorophyll ; some with a confused green mass ; some with green or brown rounded bodies of definite shape, the **zygospores**; some empty.

Under high power, notice

4. The general shape of the cells as influenced by the cell contents.

5. The *conjugating tube :* note

a. The enlargement at the middle, where an indentation marks the line of union of the two originally separate portions.

b. In some cells which have not yet conjugated, a greater or less protuberance on the side next the accompanying filament ; the beginning of a conjugating tube.

6. The *cell contents.*

a. By studying various specimens, trace the changes
    from the vegetative condition. through the several
    stages of disintegration of the chlorophyll band
    and contraction of the protoplasm to the forma-
    tion of a rounded uniformly greenish-brown mass ;
    noticing at the same time, that this change takes
    place side by side with the formation of the conju-
    gating tube. In general all the stages are easily
    found.

b. Where the conjugating tube is fully formed, note
    that one cell is empty, and the connected cell con-
    tains a single mass, the spore produced by the
    conjugation.

7. The *mature zygospore :* note

a. The *shape* and *color*.

b. The *contents.*

c. The *wall* of greater or less thickness, usually
    resolvable into two or more layers of different
    colors.

8. Make drawings to illustrate the parts and changes of
   the fruiting filaments.

ANNOTATIONS.

In the form and manner of growth of Spirogyra, we
meet with no features not seen in Oscillaria or Protococcus, except the arrangement of the protoplasm and
chlorophyll bodies. The filaments are built on the
plan of Oscillaria, with the cells larger, and the sheath
so much reduced that it can be demonstrated only with
difficulty. In some species of the closely related genus
Zygnema, however, the sheath is readily discernible.
The increase in the number of cells is effected in the

same manner as in Oscillaria, *i. e.* by the division of
the cell into halves by a transverse partition always in the
same direction, with subsequent expansion of the new
cells.

The disposition of the protoplasm shows a marked
advancement over the lower plants. Instead of being
diffused evenly through the cell, it forms a layer lining
the cell-wall, known to older botanists as the primor-
dial utricle,[5] while it only partly occupies the central
portion of the cell. The remaining space is filled by
the cell-sap, which consists of water holding various
substances in solution. The nucleus and nucleolus,
particularly the latter, are remarkably large. In the
chlorophyll band we have a unique feature; for while
it is common to have the chlorophyll separated in well
defined bodies, it is only in Spirogyra and its close rela-
tives that it assumes such peculiar and beautiful shapes.

The presence of starch granules in the chlorophyll
bodies is a very significant fact in the physiological
study of plants. They, or very similar substances,
are the first products of assimilation,[6] being the material
from which the elaborate frame-work of the plant is
eventually constructed. Usually the starch when first
formed is scattered irregularly through the chlorophyll
bodies; in Spirogyra, however, the principal part is
collected in a layer of granules about definite centers
forming hollow spheres. Within these spheres is a
highly refractive protoplasmic body, the pyrenoid.

---

[5] So named by H. v. Mohl, Bot. Zeit., 1844, p. 273; The Vegetable
Cell, p. 36.

[6] Cf. Sachs, Handbuch d. Exper.-Phys., p. 307; Textbook of Botany,
2nd Eng. ed., p. 703; Bessey, Botany, p. 178.

The starch is imbedded in the chlorophyll bodies, and is quite distinct from the pyrenoid, although the constancy in the relative position of the two would indicate some connecting influence. The pyrenoids have been long known and variously interpreted,[1] but the recent careful studies of Schmitz[8] show that they are quite analogous to nucleoli, especially in chemical constitution and mode of multiplication. They are only found in some of the algæ and in a few higher plants.

It is when we examine the fruiting of Spirogyra, that its great advancement beyond the simple forms of the protophytes becomes apparent. We meet at once with a true sexual process, which although very simple is yet clearly defined and easily traced. This process, as indeed in all other instances however modified, consists essentially of the intimate union of the protoplasm (especially of the nucleus[9]) of one cell with that of another, which after a longer or shorter period results in the production of a new individual. Usually in higher groups there is a marked difference in size, and we may conclude in other less apparent respects, between the protoplasm which is fertilized, the female element, and the protoplasm which fertilizes it, the male element. In Spirogyra a slight difference between the two elements, especially in size, has been pointed out by DeBary,[10] Wittrock,[11] and more fully by Ben-

---

[1] Hofmeister in Die Lehre von der Pflanzenzelle (1867), p. 370, calls them vacuoles.

[8] Die Chromatophoren der Algen (1882), p. 37 et seq.; Quart. Jour. Micr. Sci., xxiv, p. 246.

[9] Cf. Strasburger, Neue Untersuchungen, p. 80.

[10] Untersuchungen über die Familie der Conjugaten, 1858, p. 4.

[11] Quart. Jour. Micr. Sci., 1873, p. 123.

nett." According to Bessey,[13] however, we should consider this case the simplest kind of sexuality, in which there is as yet no differentiation into proper male and female. For the further discussion of sexuality in plants, the student is referred to the writings of Pringsheim,[14] Sachs,[15] Ward,[16] Strasburger,[17] and others.

The two plants previously examined may be found in any month of the year, but the one now under examination dies, and entirely disappears from sight by the time winter has fairly set in. It is reproduced the coming spring by the germination of the zygospores, which lie at the bottom of the water during the winter. These resting spores are admirably fitted for spanning this unfavorable season for vegetation. As a rule they require a long period of rest before reaching the germinating condition, so that while they are formed in the earlier part of the warm season, it is usually not till the following spring that they show a disposition to grow; they are dense and heavy, and therefore sink to the bottom as soon as set free by the decomposition of the filaments in which they grew; and lastly, their thick double or triple covering serves as an ample protection to the living protoplasm within.

---

[12] Jour. Linn. Soc., xx (1884), p. 430; Amer. Nat., xvii (1884), p. 421.
[13] Amer. Nat., xix (1885), p. 995.
[14] Monatsber. d. k. Akad. der Wiss. in Berlin, 1869.
[15] Textbook of Botany, 2nd Eng. ed., p. 986.
[16] Quart. Jour. Micr. Sci., 1884, p. 262.
[17] Op. cit.

WHITE RUST.

Cystopus candidus Lév.

PRELIMINARY.

THIS is a very common parasitic fungus, forming white patches on the surface of the leaves, stems and flowers of many cruciferous plants, such as various species of *Capsella, Sisymbrium, Lepidium, Nasturtium, Sinapis,* and *Raphanus.* It is especially abundant upon Capsella or shepherd's purse,[1] from early spring till late in the fall, whitening and distorting the stems, leaves and flowers. Yet, notwithstanding such luxuriant growth, the sexual condition with resting spores is not abundantly found on this host, but is, however, produced in great luxuriance inside the flowers and flowering branches of radish (*Raphanus*), causing them to become enormously enlarged, sometimes even two to five centimeters (one or two inches) across (see fig. 3).

It is possible, with patience and care, to make out the parts without the use of chlor-iodide of zinc, but it affords so much assistance that it ought to be used if obtainable.

The requisites for the following study are branches

---

[1] For a description of shepherd's purse see p. 222.

of Capsella bearing the rust, dried or fresh; the same, together with some young terminal portions of affected branches, preserved in alcohol; the swollen flowers of radish or Capsella taken when not too young, but still tender and brittle, preserved in alcohol; freshly gathered branches of rusted Capsella, or some which have not been gathered more than twenty-four hours and kept in a moist bell jar; chlor-iodide of zinc; potassic hydrate; and iodine.

LABORATORY WORK.

GROSS ANATOMY.

1. The *vegetative body* of the plant consists of delicate transparent threads, ramifying through the tissues of the host on which it grows, and can not be detected without the aid of the compound microscope.

2. The sori : in a fresh or dried specimen notice
   a. The white blister-like pustules on the surface of the host, *sori ; shape* and *extent.*
   b. The thin *external membrane,* at first entire, then becoming ruptured in the middle.
   c. The white powdery spores, **conidia,** which drop out upon jarring, if the specimen is dry.

3. Mount a section from an alcoholic specimen of radish flower containing Cystopus, stain with chlor-iodide of zinc, and notice
   a. The numerous dots scattered through the tissue of the radish, the oospores or resting spores. The staining shows them as red dots lying in a blue or yellow ground tissue.

MINUTE ANATOMY.

Mount a transverse section of an alcoholic specimen of a stem or leaf bearing Cystopus, and under low power notice

1. A layer of short vertical filaments, conidiophores,[2] together forming the hymenium, which appear to arise from the tissues of the host and bear on their free extremities

2. Chains of rounded *conidia*, now mostly detached.

The vegetative portion of the plant, consisting of branching filaments pervading the tissues of the host, can rarely be made out even after staining, without specially skillful manipulation.

3. The *everted membrane* formed from the surface cells of the host, formerly covering the sorus.

4. Draw.

Under high power notice

5. The *conidia* : exact shape, wall and contents.

6. The delicate neck or *pedicel* supporting each conidium before becoming detached.

7. Draw a conidiophore with its conidia.

Take a piece of the host bearing conidia and boil for a minute or two in potassic hydrate ; remove a portion to the slide, tease apart thoroughly with needles, and stain with chlor-iodide of zinc. Notice

8. Much branched, often matted filaments, mycelium, pulled out from the tissues of the host.

---

[2] Cf. fig. 8.

a.　The irregular thickness of the mycelial filaments, or **hyphæ.**[3]

b.　The absence of transverse partition walls.

c.　Draw a few hyphæ.

9.　The groups of conidiophores.

a.　The manner in which the conidiophores arise from the vegetative hyphæ.

b.　The successive degrees of abstriction of the conidiophores resulting in the formation of the spores.

c.　Draw a group of conidiophores with the attached hyphæ.

Prepare a slide as before, using the immature terminal part of the branch bearing the Cystopus, preferably a flowering branch ; **search** among the untorn tissues of the youngest organs, particularly in the pedicels of the young buds, for the extremities of the advancing hyphæ.[4] After noting the **more direct** course of the hyphæ, and the fewer branches, observe

10.　Very small globular bodies lying along the side of the hyphæ, **haustoria** or sucking organs.[5] They usually appear brighter than the hyphæ, and are quite abundant.　If the illumination is sufficiently strong, observe

a.　The very delicate *stalks* by which the haustoria are connected with the hyphæ.

b.　Draw some hyphæ with haustoria.

---

[3] *Hypha* is the name applied to a single filament, while *mycelium* is a collective term for a number of hyphæ.

[4] If properly stained there will be no difficulty in distinguishing the mycelium from the tissues of the host.

[5] It is difficult to demonstrate these without proper staining.

Dust some conidia from a fresh growing plant[6] upon a slide and mount with water ;[7] in about an hour, notice

11. The *small protuberance* formed on one side of some of the conidia, which **opens** and permits the escape **of** the protoplasm **in the form of several** motile bodies, **zoospores.**

a. The *shape* of the zoospores, and **the pair of** bright spots in each.

b. Study the *movement.*

c. Notice the pair of delicate vibratile **cilia,** by means of which the movements are effected. Stain **with iodine,** and the cilia can **be** seen more easily. Note their *length.*

d. The *color* imparted to the zoospore and its cilia **by the iodine.**

e. Draw some zoospores, and also one or two conidia which have not discharged **zoospores, and one or** two empty **ones.**

12. The *sexual reproduction.* **Stain a section of** an alcoholic specimen of radish flower containing oospores **with** chlor-iodide of zinc, and notice

a. The numerous globular bodies, stained wine-red, lying in the tissues of the radish, **oogonia.**

b. **Accompanying** them, and stained the same, smaller **rounded bodies, antheridia.**

c. **In some of the** oogonia, a globular mass of granular **protoplasm,** not completely filling the oogonium, the **oosphere.**

d. A slender **tube** passing from the antheridium to

---

[6] The conidia will germinate if sown at any time of day, provided the specimens are fresh, but will do so more readily when sown in the morning from plants which have remained over night under a moist bell jar.

[7] Care must be taken that the water does not evaporate, and to guard against this it is best to use a slide having a shallow cell.

the oosphere, the **fertilizing tube** ; very difficult to demonstrate. Draw.

e. In older oogonia, more opaque roughened bodies the **oospores**, formed from the oospheres. Note
  i. The *flexuous ridges* on the exterior.
  ii. The *contents*, in spores not too mature.

f. Draw some oogonia and accompanying antheridia showing different stages of development of the oosphere and oospore.

Tease out some tissue containing oospores, which has been boiled in potassic hydrate, stain lightly or not at all, and notice

g. The manner in which the oogonia and antheridia arise from the vegetative hyphæ. Draw a few examples.

h. The rather strong, pointed *beak* sometimes to be seen on one side the antheridium, the fertilizing tube which has been pulled out of an oogonium. Draw.

ANNOTATIONS.

In Cystopus we have a much simplified condition of an advanced type of development. The higher development is shown in its sexual reproductive apparatus, the sexual elements being quite dissimilar in size and in behavior. The larger (female) element, the oogonium, receives the protoplasm of the smaller (male) element, the antheridium, the former remaining in a passive state, while the antheridium is the active agent in securing the union. This is the essential plan for all higher plants, as well as for the group to which Cystopus belongs, the Oophyta.[8] The transfer of the

---

[8] The terms Zygophyta, Oophyta and Carpophyta are used for the three great groups of lower plants, in accordance with the suggestion of Prof. C. E. Bessey in the American Naturalist, xvi (1882), p. 46, and first introduced in his Essentials of Botany, 1884.

protoplasm by means of a fertilizing tube, and the subsequent formation of a thick-walled resting spore is very similar to what takes place in Spirogyra. In both cases the spore clothes itself with a thin inner wall, very difficult to see clearly, and an outer, thick protective wall. In Cystopus this outer wall is sculptured in a manner characteristic of the species. The oospores thus formed remain over winter ; the tissues in which they lie become disintegrated ; they are distributed by rain and wind, and finally germinate.

Next to the mode of sexual reproduction, the most interesting feature about the plant under consideration is its habit of life and the adaptations which have been induced thereby. It is throughout its existence a complete parasite, growing and feeding upon plants of a very high degree of organization. Being no longer required to elaborate food for itself, finding it always at hand and of superior quality, it possesses no chlorophyll bodies by which it might assimilate its own food, and is therefore quite colorless. As it grows, it sends its branching filaments ramifying throughout all the softer tissues of the host. They do not penetrate the cells, however, but push about between them, and in order to extract the nourishing fluids readily, especially in the newest portions where rapid growth is taking place, send out sucking tubes or haustoria, which penetrating the adjacent cells expand into minute absorbing bulbs.

The means of distribution which the plant possesses in its oospores is rather limited, being inferior to that of Spirogyra ; and when once established in a host it is debarred from all further locomotion, such as the

moving water imparts to the spores of Spirogyra. In
order to secure certain and extensive distribution, therefore, and to provide for a succession of crops through the
growing season, it produces conidia or summer spores
in the greatest profusion, which being light and dry
are easily blown about by the wind, and are ready to
germinate at once. The thin wall and active protoplasm of the conidia, from which they derive this
advantage, render them at the same time short lived,
so that if a conidium does not find favorable conditions for growth within a few hours after reaching
maturity, it perishes. The conidia germinate in water,
and with best results in a film of water, such as is
formed by heavy dew. To still further promote distribution, each conidium breaks up into several active
zoospores, which, after moving about for fifteen minutes or so and finally coming to rest, put out a mycelial tube that penetrates the host, and forms a new
plant. The zoospores, except in being colorless like
the parent, remind us of those of Protococcus, serving
the same purpose of distribution and reproduction.

The absence of septa, except for the separation of
the antheridia, oogonia and conidia, making the vegetative portion a continuous cavity, is a character
shared with many other members of the Oophyta, both
colorless and green forms, and with some of the molds
belonging to the Zygophyta.

The student has doubtless been struck with the
rarity of the cases in which he could detect a fertilizing
tube, even where the antheridium appeared to lie in
the proper plane. There is doubtless a reason for this
aside from the mere difficulty of manipulation, which

is to be sought in the nature of the parasitism exhibited by Cystopus. Whatever may be the full significance of sexuality, many facts point to the belief that it is an expedient for the reinvigoration of the exhausted energies of the plant.[9] Cystopus is intimately associated with a plant immensely above it in the scale of development and of a correspondingly higher potentiality. Its vigor is in direct ratio to that of its host, which latter far exceeds the requirements of the simple parasite. The energy which the parasite receives from its host may take the place to some extent of that usually obtained through the sexual process. It therefore seems justifiable to believe that while the antheridia are in most cases formed, the fertilizing tube is often either not present or functionless, *i. e.* that we have the production of oospores without the aid of the male element, a method known as parthenogenesis,[10] a difficult matter to establish by observation. This view is rendered more probable by the fact that it is the customary mode of reproduction in some of the closely allied Saprolegniæ[11] which are mostly parasitic for at least a part of their life upon insects, a still more highly organized food than that obtained by Cystopus and its immediate allies.

[9] Ward, Quart. Jour. Micr. Sci. xxiv, (1884), p. 303; Bot. Gaz. ix, p. 146.

[10] Sachs, Text-book of Botany, 2nd Eng. ed., p. 902; Ward, l. c., p. 307.

[11] Pringsheim in Jahrb. f. Wiss. Bot., ix; DeBary, Beiträge zur Morph. u. Phys. der Pilze, 4te Reihe, p. 73; Sachs, l. c., p. 275.

THE LILAC MILDEW.

Microsphæra Friesii Lév.

PRELIMINARY.

THE mildew on lilac is extremely common in the United States, making the upper surface of the leaves look white and moldy from midsummer on. The first stage at which the fungus is ready to gather is when it appears powdery, which is usually in June or July, the earlier collections being the best. The next gathering should be made in the early part of September, and another just before the leaves fall. As the leaves bearing the fungus are gathered, lay them in a book or plant-press to dry. If it is possible to examine the first stage with fresh material it will prove more satisfactory, but for the remainder dried material will answer quite as well.

The required material consists of dried lilac leaves bearing the fungus, gathered in midsummer and autumn ; and potassic hydrate.

LABORATORY WORK.

GROSS ANATOMY.

A. GENERAL CHARACTERS. Notice

1. The *distribution of the fungus* over the surface of the leaf.

2. The *color*.

B. THE CONIDIA. Notice

1. The pulverulent appearance on the leaves first **gathered,** caused by the abundant *conidia*.

C. THE FRUIT. Notice

1. **The black** dots on leaves gathered later in **the season,** the spore-fruits or **perithecia.**

2. Associated with the black dots, other yellow ones, the *immature fruits*.

MINUTE ANATOMY.

A. THE MYCELIUM. Scrape the fungus from the surface of a leaf gathered in early summer, having first moistened it with potassic hydrate if the specimen **is a** dried one, and under high power notice

1. The colorless *filaments* of the mycelium.
   a. The *branching*.
   b. The irregular *diameter*.
   c. The rarity of *partition walls*.

2. Small lateral expansions of the filaments, *haustoria*, somewhat like irregularly indented disks with very **short** thick stalks. Generally difficult to find.

3. **Draw.**

B. THE CONIDIA. Prepare a slide as before from **a** pulverulent surface, and notice

1. The abundant *conidia*, separated and free, owing to the manipulation.
   a. Their *shape* and *color*.
   b. **The** *cell-wall* and **contents.**

2. **The** conidia-bearing branches, or *conidiophores*, **which** leave **the** mycelial filaments at **right** angles, and **are** provided with cross partitions at regular intervals, **and**

to which may yet be attached some fully formed spores.

3. Draw some conidiophores and conidia.

C. **THE** PERITHECIA.   Prepare a slide as before with mature fruit, and **notice**

1. The *shape* and *color*.

2. The *reticulations* of the surface due to the cellular structure.

3. The **appendages** extending out from the sides.   Note
    a.   The *number*.
    b.   The *color*.
    c.   The *length* compared with the diameter of the ` perithecium.
    d.   The *cross partitions*, if any.
    e.   The manner of *branching*, and the number of times in each.

4. Draw a perithecium with its appendages.

Crush the perithecia while watching them through the instrument, by pressing on the cover-glass with a dissecting needle, and notice

5. The escape of sacs containing spores, **asci**.   Note
    a.   The *number* from each perithecium.
    b.   The general *shape*.
    c.   The short *pedicel* or beak by which they were attached.
    d.   The thin part of the wall at the *apex*, not to be seen in every case.
    e.   The number of spores (*ascospores*) in each ; their shape.
    f.   Draw an ascus with its spores.

6. Examine younger and younger perithecia to as early **a** stage as possible. **Draw.**

D. The **very** simple ORGANS **OF** FERTILIZATION, the beginning of the perithecia, **can** rarely **be found**[1] ; if **seen,** notice

   a.   The larger axial cell, the **carpogonium.**

   b.   The smaller lateral cell, applied closely to the car-pogonium, the *antheridium.*

   c.   Draw.

ANNOTATIONS.

The **group of** Carpophyta to which **Microsphæra** belongs, a very large one, is characterized **by having a** special covering for the spores, developed **as a result of** fertilization.   Except in some of the higher forms, **the** fertilization takes place much as in the Oophyta, **but the** subsequent development **is very** different, **for an out-**growth **of** branches **from** the portion immediately below the organs of fertilization **at** once arises which eventually envelops the forming **spores and** develops **into the body** of the fruit.

  It is altogether likely that Microsphæra has reached **an advanced** parthenogenetic stage, *i. e.* the fruits **are** largely **produced** without the transfer of protoplasm from the antheridium **to the** carpogonium, which **consti-**tutes fertilization.   **On this** account **some other** plants better illustrate the fertilization and **the early** growth **of** the fruits than the **one used** ; the **student** is advised **to examine** these features, **if possible, in** *Nemalion,* one

---

[1] To get some idea of their shape, **examine figs. 188 and 189 in Bessey's** Botany, p. 280–1.

of the marine algæ, or *Batrachospermum,* one of the fresh-water algæ.

The comparison of Microsphæra with Cystopus is very instructive in showing how practically the same ends have been reached by widely different plants. Both are parasitic, the one living inside the host, the other upon its surface, both deriving nourishment by means of haustoria, in addition to what is absorbed directly through the walls of the filaments. It is somewhat doubtful, however, if the haustoria of Microsphæra pierce the cells of its host, although those of some closely related species are thought to do so.[2] Both bear aerial asexual spores, which are formed by successive abstrictions from vertical mycelial threads, the main difference being that in Cystopus these must break through the surface tissue of the host, and are therefore required to grow in groups in order to exert the necessary force, while in the superficial Microsphæra they are single, and evenly distributed. The conidia of Cystopus germinate by formation of zoospores, while those of Microsphæra grow a mycelial filament at once, a difference due to some obscure cause. Both plants form resting spores, but in Cystopus the protective covering is the thickened wall of the spore, in Microsphæra it is a specially developed shell, inclosing a number of spores in sacs.

There is not much known of the manner in which these fruits pass the winter and give rise in the spring to another growth of mildew.[3]   It is plain from

---

[2] Cf. Bessey, **Botany,** p. 279.

[3] Wolff has studied the germination of the ascospores in *Erysiphe graminis.* Bot. Zeit. 1874, p. 183.

the structure, however, that the spores escape from the sacs through the thin spot at their apex, but not so evident how they escape from the shell of the fruit and reach the host plant. The appendages we may suppose are of some service in distributing the fruits.

COMMON LIVERWORT.

Marchantia polymorpha **L.**

PRELIMINARY.

THIS plant is common throughout America and Europe. It grows among grass, over wet soil or rocks, in dryer spots along walls and fences, and occasionally in more exposed situations, but is most luxuriant in damp shady places. The vegetative part consists of flat, green, leaf-like stems, twelve millimeters (half inch) or so wide and five to eight centimeters (two or three inches) long, appressed to the ground, held down by numerous silky hairs on the under side, and much branched, usually forming extended mats.

There are two sorts of reproductive branches which occur on separate plants. These branches (see fig. 2) are slender stalks about an inch high, bearing flat disk-like heads a quarter of an inch or more across—the male with scallops, the female with finger-shaped rays. The two forms sometimes grow at the same spot or locality, but quite as often entirely apart from each other. Besides these organs there are often small sessile cups (cupules) on the upper surface of the stems, containing green grains.

If either cupules or reproductive branches are present,

no other plant is likely to be mistaken for it. In their absence it may be told from any of the lichens by the small, diamond-shaped markings on its upper surface. A common liverwort growing in damp places (*Conocephalus conicus*) may be known in its sterile condition by its larger size, larger areolæ and more prominent stomata which in Marchantia are barely visible to the naked eye, but in *Conocephalus* are as large as pinholes. A common greenhouse liverwort of similar appearance (*Lunularia cruciata*) may be distinguished by its crescent-shaped cupules, lacking a border on one side.

Marchantia grows luxuriantly in the greenhouse, producing an abundance of cupules, and often fruiting. It may be placed on the pots in which other plants are grown or given a bed to itself.

When gathering material, care should be taken to save fertile plants with young heads; in female plants, especially, some heads should be no larger than a quarter the size of a pinhead, and which at this stage of growth are to be detected in the sinus at the growing end of the stem.

To complete the laboratory work requires fresh or alcoholic specimens bearing cupules and both kinds of reproductive branches; fresh specimens of the male and female heads; and iodine.

LABORATORY WORK.

GROSS ANATOMY.

A. GENERAL CHARACTERS. Note

1. The flattened horizontal stem or thallus, composing the larger part of the plant.

a. The *branches ;* all lying in the same plane as the main axis, except

b. The fruiting branches, consisting of erect stalks, pedicels, supporting disk-like heads or receptacles of two sorts, to be found on separate plants :

  i. The **antheridial** (sterile) with scalloped heads, and

  ii. The **archegonial** (fertile) with star - shaped heads.

2. The numerous *hairs* on the under surface of the thallus.

3. The dark brown or purple **leaves**, somewhat concealed by the hairs, and closely overlapping to form a low ridge along the median line beneath.

4. The **scales** along the sides of the thallus beneath, some projecting beyond the margin ; more conspicuous on plants grown in damp shady places.

5. Sessile cups or **cupules** (very prominent when present), seated on the upper surface of the thallus, containing bright green flat bodies, the **gemmæ**.

B. THE STEM. Note

1. The *color* of the upper and lower surfaces in fresh specimens.

2. The well marked median line, **midrib** ; and the broad expansions, **wings,** on either side of it.

3. The indented *apex.*

4. Mode of branching, **dichotomous** ; each stem is resolved into two equally diverging stems, one of which soon exceeds the other by more rapid growth, giving the false appearance of being monopodial.

5. On the upper surface the small areas, **areolæ,** best seen on the older parts, in the center of each of which is

    a.   A circular breathing-pore, or **stoma**, readily detected with the lens.[1]

6. On the under surface, notice **the absence of stomata** and areolæ.

7. Make an outline sketch of a branching stem **to show** the contour, the median line, and the mode of branch-**ing.**

8. Mount a transverse section, and if from a growing plant, notice the pale *middle tissue*, the green *upper surface*, the dark-colored *lower surface*, and the group of *median leaves* projecting downward from the midrib ; **if** from an alcoholic specimen the color is wanting. If **the** specimen is very thin and carefully prepared, long narrow *air-cavities* may be seen just beneath the upper surface with possibly stomata leading out from the center of some.

C. **THE LEAF.** Remove **the hairs from the lower sur-**face **of** the stem, and notice

  1. The direction and manner **of the** overlapping of the *leaves.*

  2. The *shape.*

  3. **The** curvature and extent of the line of *insertion.*

  4. **Illustrate shape, and** position on the stem by diagrams.

D. **THE** TRICHOMES. These are **of** two kinds, **the** hairs and the scales.

  1. The *hairs.* Notice

    a.   The silky mass extending downward along the midrib, serving for roots, **rhizoids** : the part of the thallus from which **they** arise.

---

[1] The areolæ and stomata are very large in *Conocephalus conicus*, the latter being plainly seen without the aid of a lens.

    b.   The **closely** appressed *strengthening hairs* on the wings; **their** origin and direction of growth.

  2. The *scales.* Notice

    a.   The slightly projecting *marginal scales*, along the under **edge of the** thallus ; insertion and direction.

    b.   The colorless *intermediate scales*, **midway** between **the** edge and the midrib ; **insertion** and direction.

  Mount both kinds of scales and notice

    c.   The *shape* of each. **Draw.**

E. THE CUPULE. Note

  1. *Position* on the **stem.**

  2. *Shape* and *size.*

  3. The degree of *smoothness* **of the** outer and inner surfaces.

  4. The thin *margin,* infolded **when** young ; **shape and** regularity of the *teeth.*

  5. The *gemmæ* within ; **remove** some **and place on** a **white surface,** and **note** their form—usually two **notches can be** detected opposite **each other.**

  6. **Draw a cupule with** its gemmæ.

F. THE FRUITING BRANCHES.

  1. *Position* on **the stem** ; note that **they are** always continuations **of the** midrib, **and** consequently **apical,** although **sometimes,** by the prolongation of **the wings,** appearing to **rise** from the upper surface.

  2. The *pedicel ;* notice

    a. *Color* and *striation.*

b. The flat, green, *posterior*[2] *surface* of the archegonial branch, wanting in the antheridial.

c. Pull a pedicel in two and notice the hairs projecting from a pair of *grooves* on its anterior face.

d. Make a transverse section of the antheridial pedicel and notice the *outline*, and the position and form of the grooves. Draw.

e. In a similar section of the archegonial pedicel notice the outline, grooves, and the posterior *chlorophyll-bearing portion*, with its row of air cavities. Draw.

3. The *head of the antheridial branch*, consisting of a large receptacle and minute and inconspicuous antheridia imbedded in it ; notice

a. The *general shape* of the receptacle.

b. The particular *outline* of the margin.

c. The broad radiating *ridges* on the upper and lower surfaces.

d. The narrow wing-like *margin*, more easily distinguished by holding the head toward the light.

e. The numerous *scales* on the ridges beneath, most abundant toward the margin.

f. Cut a vertical section and observe the rather large oval *cavities* beneath the ridges ; each cavity contains a single sac, *antheridium*, holding the innumerable fertilizing bodies, antherozoids,[3] neither distinctly visible. If the antheridium is still

---

[2] I. e., the surface looking away from the axis of the stem and corresponding to its upper surface.

[3] The antherozoids are far too small to be seen except with a compound microscope ; they escape through openings in the upper surface, also too small to be made out in this connection.

distended with antherozoids, it will completely fill the cavity and appear as a darker or lighter spot in the tissues, according to the thickness of the section, but if the antherozoids have escaped, the collapsed antheridium remains, although it can rarely be detected, and the cavity appears empty. The antherozoids may sometimes be seen in mass as a faint cloud escaping into the water of the slide, especially when pressure is applied to the cover-glass.

g. Make a horizontal section from the upper part of the head, after first removing a thin surface slice, and again observe the cavities.

h. Draw an uninjured antheridial branch, giving prominence to the upper surface of the head.

4. The *head of the archegonial branch*, consisting of a star-shaped receptacle and circle of reproductive organs beneath ; notice

a. The *receptacle* ; its *general shape.*
   i. The *rays*, with a longitudinal crease beneath ; their number.
   ii. The *cleft*, which extends to the posterior side of the pedicel.

b. The *reproductive organs* forming groups alternating with the rays.

c. Carefully separate one of the groups with a needle, without detaching it, and notice
   i. The border, **perichætium**, surrounding it, and inclosing
   ii. The several young **sporogonia**. With a needle remove the sporogonia to a slide without injuring the perichætium. Now observe that
   iii. The two halves of the perichætium are united

at an acute angle next the pedicel, and by an infolded flap next the edge of the receptacle. This flap is best seen by spreading apart the rays between which the perichætium is situated.

iv. Remove the perichætium and spread out on the slide with the sporogonia already placed there ; notice

  α. The fimbriated free edge of the *perichætium*.

  β. The opaque *sporogonia* with their very short thick stalks, each inclosed by a delicate sheath, the perianth, twice the length of the immature sporogonia, but equaling or even shorter than the older ones. Draw.

v. Tear away the perianth and notice that it is quite free from the sporogonium, which, with its stalk, can now be seen more clearly. Draw.

d. In a head from a fresh specimen[4] having mature sporogonia protruding from the perichætia, notice

i. The *form* of the sporogonia.

ii. The *contents ;* a fluffy yellow mass when escaping from a freshly ruptured fruit.

iii. Mount and notice the dust-like part, *spores*, and the short delicate hairs, elaters.

iv. Breathe gently upon a mass of dry elaters and notice the *movement* caused by the moisture of the breath.

e. Draw an uninjured archegonial branch, giving prominence to the under surface of the head.

---

[4] An alcoholic specimen does as well, except to illustrate ii and iv infra.

MINUTE ANATOMY.

A. **THE** TRICHOMES. Remove some of the hairs that are pressed tightly upon the **under** surface **of the** wings of the stem, and under high power notice

1.　The *shape*.

2.　The *internal projections*, **sometimes horn-like**, branched, and **extending quite across the cavity of the hair.**

3.　**The more or** less prominent *spiral constriction* on which **the projections are** seated, giving **the walls in** optical **section the** appearance of alternate **scallops.**

4.　**Draw.**

5.　**Compare the** loose silky hairs along the middle **of the** stem, the *true rhizoids*, with the preceding.　**Draw.**

Mount marginal and intermediate *scales*, **and notice**

6. The cellular *structure* of each.　Draw.

B.　THE STEM OR THALLUS.　Remove **a thin slice** from the upper surface of the stem, mount with **the free** surface uppermost, and under high power notice

1.　The surface or epidermal **cells; their shape and** con-tents.

2.　**The large** breathing **pores or** *stomata*, encircled by **rows of** special cells, the **supplementary guard-cells;** **number of cells in each** circle.

3.　**Draw a** stoma with some surrounding **tissue.**

Remove the cover-glass, remount the **section with the** **free surface** downward, and disregarding **other features,** notice

4.　The four innermost (now lying uppermost) cells of the stomata, having more or less **prominent projections** (sometimes obsolete) arching over toward **the center** of the pore, the **true or active guard-cells.　Draw.**

Take a **very** thin slice **in the** same way from the lower surface, and notice

5. The shape of the cells of **the** *epidermis*, **absence of** stomata, and the insertion of the *hairs*. **Draw.**

Make a vertical section **of the** stem **parallel to the direction of** the radiating tissues of the wings, which **is usually at about** an angle of forty-five degrees to the **line of the** midrib ; notice

6. The wider colorless *under part* of the **stem.**

7. The narrower chlorophyll-bearing *upper part.*

In the colorless portion of the stem, notice

8. **The** closely packed **parenchyma** cells, bordered **on one side** by the marginal **row** of cells **forming the** *lower epidermis.*

9. **The shape of** the parenchyma cells, their uniformity **of size, and the transverse** *reticulated thickenings.*

10. **The** smaller *size* **of the** epidermal cells, **forming an** indefinite marginal **row,** the *walls* plain, **and** either colorless, purple **or brown.**

11. **Draw** several cells of the lower epidermis, and some **of** the adjacent parenchyma.

In the upper green layer **of the** stem, notice

12. The large *air cavities*, from the **bottom of** which **a** thick growth **of (in fresh** specimens very green) cells arises, branching **in** a cactus-like manner.

13. Note the **shape of these** cells, their manner of union, **and the** rounded **(in** fresh specimens bright green) **chlorophyll grains. Draw.**

In **the** same section, **if** sufficiently thin and perfect, ex- amine

14. The *partition wall* which separates contiguous cavities,

and the over-arching *roof* formed of the single epidermal layer of cells. Draw.

When a section is found which has passed through a *stoma*, notice

15. The chimney-like *structure*, the number of cells in depth, and the shape of the cut ends of the cells, especially of the outer and innermost ones. Draw.

16. Illustrate the structure of the stem as shown in the longitudinal section by a diagram.

17. Cut a vertical section of the wing at right angles to the longitudinal one already examined, and compare with it : especially note the difference in the outline of the parenchyma cells, and the frequent absence of reticulated markings. Draw.

C. THE LEAF. Remove a leaf from the stem, best taken from a young stem where the leaves are comparatively large and conspicuous, and notice

1. The shape of the *cells*, absence of markings and of chlorophyll, and the uniformity of the cells throughout the leaf. Draw.

Make a transverse section of the stem at right angles to the midrib, and a good *transverse section of the leaves* will usually be obtained : notice

2. The single row of cells forming the blade, or sometimes two or three rows at the base, and the manner in which the older leaves over-arch the younger. Draw.

D. THE CUPULE. Remove a cupule and place face upward on the slide, ignoring for the present the gemmæ which float out from it ; press the cover-glass down until the cupule is sufficiently flattened out, when it will appear as a wide ring of tissue, the bottom of the cup having been cut away. Examine with low power

1. The *border* of triangular teeth ; the fimbriated sides and elongated apex of each tooth. The inner **part of** the cupule is too thick to be seen well.

    Place under a higher power.

2. Examine the structure of the *teeth* and their marginal hairs more carefully, and draw. Vary the focus, and notice whether the inner surface of the cup is smooth or rough.

    Remove **the** cover-glass, turn the specimen over, and ex-**amine as** before. Notice that

3. The outer surface of the cup is covered with short hairs or *papillæ*.

    Make a vertical section passing through the center of **a** cupule and through the stem on which it is seated, examine under low **power,** and note

4. **The two parts of** the *cupule.*
    a. **The** *base.*
    b. The abruptly spreading *limb.*
    c. Arising **from** the bottom of the cup, flattened tri-chomes, **the** *gemmæ,* in various **stages of develop-ment.**
    d. Illustrate with diagram.

    Examine the *limb* under high power, noting

5. The small cells of the inner epidermis, the parenchy-matous tissue beneath, the absence of **a** distinctly **marked outer** epidermis, and the short one- or two-celled hairs on **the outer surface.** Draw.

6. Examine next **the** *base* **of** the cupule, noting the **man-**ner in **which** the layer of air cavities of the stem and their chlorophyll tissue is continued up the outside of **the** cup as far as the insertion of the limb.

7. **Upon** the inside of the cup, observe
    a. The numerous glandular, one-celled *hairs,* and **among them**

   b.   **Thicker hairs** in various stages of cell-multiplication, from the first division into two cells by a transverse wall, to the fully formed many-celled *gemma* still attached by its *pedicel*.

   c.   **Draw** part of the bottom of the cup, to show insertion and form of the glandular hairs and young gemmæ.

8.   Examine under a low power the *mature gemmæ* which have floated from the cupule, and note

   a.   The *shape.*

   b.   The cellular *structure.*

   c.   Cells here and there devoid of chlorophyll.

   d.   The *scar* left by the pedicel.

   e.   The pair of *vegetative notches* placed midway, one on the right, the other on the left side, in which, when the gemmæ are sufficiently mature, may be seen

   f.   The early stage of the new *plantlets* in form of delicate papillæ.

   g.   Draw.

E. THE ANTHERIDIAL BRANCH.

1.   The *pedicel.* Pull in two a pedicel and remove some of the hairs which protrude from two grooves on its posterior face, and lay upon a slide. Now make a transverse section of the pedicel and mount with the hairs.

   Under low power, notice

   a.   The general *outline* of the section, the two conspicuous *grooves* or channels, and the uniformity of the whole tissue. Illustrate with diagram.

   Under high power, notice

   b.   The colorless dense *parenchyma* forming the mass of the pedicel, bounded by the surface row of small *epidermal cells*, in **fresh** specimens usually colored purple or brown. Draw.

c. The overlapping projections inclosing **the right and** left channels or grooves, **from the apex and sides** of which arise *leaves*, which seen in cross-section appear as a single row of cells, **or** double **rowed** at the **base ; observe the** manner of infolding, comparing sections when convenient from different parts of the same pedicel, and from different pedicels, drawing the most interesting.

d. In some **of** the grooves will be found excellent *cross-sections of the hairs*. Examine now the hairs pulled from the grooves. Draw.

e. Remove **a** thin paring from the anterior surface of the pedicel, examine the *epidermal cells*, **and, by** varying the focus, the face of the leaves. **Draw** a few cells of each.

2. The *receptacle*. Take **a** slice from **the** ridges **on the** upper surface of the receptacle, mount with **the free** surface uppermost, **and notice**

a. The *stomata* and *epidermal cells*.

b. The *pores*, around which the epidermal cells converge, the mouths to the underlying cavities containing the antheridia.

c. **Draw.**

Remount the section with the free surface downward, focus on the **cut** surface, and in the thicker part of the section notice

d. The large **air** *cavities*, producing from the sides branched *chlorophyll filaments* like those of the **stem.** Focus deeper into the cavities and notice

e. The *stomata*, the four innermost cells inflated and almost or quite closing the pore of the stoma.

f. The *pores* of **the** antheridial cavities situated in

the walls between the air cavities; also the disposition of the surrounding tissue.

Cut a rather thick vertical section a little to one side
of the center of an immature receptacle, and notice

    g.   The chlorophyll cavities,with their chlorophyll cells.

    h.   The much larger antheridial cavities, which are
quite likely to be empty, or may contain the
membranous remains of the antheridial sac, or may
be more or less filled with

3. The *antheridium with its paraphyses ;* notice

    a.   The *shape* of the antheridium.

    b.   The *pedicel* by which it is attached to the bot-
tom of the cavity.

    c.   The structure of the *wall*, brought into view by
focusing on the part nearest the eye.

    d.   The wall as seen in optical section, only a single
cell in thickness.

    e.   The uniform *contents*, consisting of very small
squarish cells, filled with colorless protoplasm.

    f.   The several unicellular paraphyses surrounding
the base of the antheridium, and not much longer
than its pedicel ; best seen when the antheridia are
young.

    g.   That the antheridia are younger toward the margin
of the head, and older toward the center.

    h.   Draw an antheridium with its paraphyses.

4. The *antherozoids ;* if the section just examined be from
a freshly gathered specimen, the contents of many of
the antheridial cells will have escaped into the water of
the slide[5] ; notice

---

[5] An excellent way to obtain antherozoids for examination is to place
a small drop of water on a slide and hold a freshly gathered head in it for
a few moments, when, if the antherozoids are ripe and abundant, they
will make the water milky.

a. The rapid *motion* of the antherozoids, becoming slower and slower until after some time they come quite to rest.

b. Their *form ;* a slender filament, at the anterior end of which may be detected, when the motion becomes sufficiently slow,

c. Two very delicate vibratile cilia; the form and motion may be more readily studied by staining with iodine, and watching the antherozoids as they pass gradually under its influence.

d. The delicate hyaline vesicle and its contents, dragged about by some of the antherozoids until finally detached.

If the section be from an alcoholic specimen, some antherozoids will have escaped, or can be made to escape by pressing on the cover-glass, when the form can be studied as before, but the filaments will be found quite closely coiled, the cilia difficult to detect, and the vesicle probably invisible.

F. THE ARCHEGONIAL BRANCH.

1. **The** *pedicel ;* in a transverse section under low **power,** notice

    a. The general *outline.*

    b. The two *grooves.*

    c. **The** *posterior plate* containing air cavities and chlorophyll tissue.

    d. Illustrate with diagram.

    Under high power, notice

    e. The larger rounded anterior part, in every essential like that of the antheridial pedicel.

    f. The smaller flattened posterior part in which lie

        i. The *air cavities,* like those of the thallus, except smaller, sparsely provided with

    ii. *Chlorophyll-bearing filaments,* springing indifferently from the floor or walls.

    iii. The (usually single) layer of small cells forming the floor, partition walls, and roof of the cavities.

    iv. The *stomata,* these will occasionally be cut across ; note the number of cells in depth, and their relative size.

    v. Make a drawing to illustrate the several points.

Make a longitudinal antero-posterior section and note

    g. The length of the *parenchyma cells,* and shape of the air cavities.

    h. Remove a thin paring from the flat posterior surface of the pedicel, and mount with the free surface uppermost, noticing the *epidermal cells* and stomata. Draw.

    i. Remount the section with the free surface downward ; note the relative size of the air cavities, and the appearance of the stomata.

Cut off a pedicel near the base, make an antero-posterior longitudinal section of the basal portion, together with the stem from which it arises, and note

    j. The continuity of the tissues of the pedicel with those of the stem.

Cut off a pedicel near the head, make an antero-posterior longitudinal section passing through the pedicel and through the cleft in the receptacle, and note

    k. The continuity of the tissues of the pedicel with those of the receptacle.

  2. The *receptacle ;* cut a transverse section of one of the *rays,* and notice

    a. The *central cavity* in which lie numerous hairs like those in the grooves of the pedicel.

    b. The *encircling tissues*, indented at **one point, yet** continuous ; notice further

        i. The internal portion of uniform parenchyma.

        ii. The external row of *air cavities*, containing chlorophyll-bearing filaments, **and** provided with stomata, essentially like **those of the** pedicel.

        iii. *Papillary trichomes* arising from many of **the** epidermal cells.

Using an immature branch, cut a transverse section across two or three rays nearer the center of the head and passing through the groups of sporogonia, notice

    c. The central cavity, much smaller than in the rays.

    d. The **right** and left sides, which instead of being **united,** are prolonged into the *perichætium,* **so** that **the** perichætial **leaf on the** right side **of the group of** sporogonia belongs **to the left** side **of** the right hand ray, while **the** perichætial leaf **on** the **left side belongs to the right side of** the left hand ray.

    e. The section of the perichætial leaves, one cell in thickness, or sometimes two at the base.

    f. Examine the flat surface of the perichætium, the shape of the cells, and the notched and *fimbriated margin.* Draw.

    g. **The** bent filaments, *paraphyses*, composed either **of a single row** of cells, or of two or more united **rows for** part or the whole length. Draw.

  3. **The archegonia,**[6] the flask-shaped **bodies among the paraphyses, consisting of**

---

[6] Bear in mind that the archegonia are called sporogonia after fertilization and a certain amount of growth has taken place.

a.   The bulbous *base :* in optical section make out a single layer of cells inclosing a central cavity
b.   The long *neck.*[7]
c.   A ring rising up around the base in some cases, the early stage of the *perianth.*
d.   Draw.

4. The *sporogonia ;* selecting the immature ones, notice under a low power

a.   The *perianth ;* its deeply notched margin, which is usually twisted over the fruit ; observe the cellular structure. Draw.
b.   Tear away the perianth, examine the surface of the sporogonium and its *stalk,* and notice the remains of the neck of the archegonium.
c.   Crush some of the sporogonia by pressing upon the cover-glass, noticing the escaping *contents* consisting of slender threads having granular protoplasm and pointed ends, the immature **elaters,** and rows of young *spores,* both radiating from the base of the fruit. Draw.
d.   Examine some *mature spores ;* notice
      i.   The *wall.*
      ii.   The *contents.*
e.   Examine the *mature elaters ;* notice
      i.   The delicate *wall,* not easily distinguished.
      ii.   The *spiral bands.*[8]
      iii. Examine some dry elaters without a cover-glass, and observe the *movements* when dampened by the breath.

---

[7] These archegonia, unless taken from a very young head, are mostly sterile, not having been fertilized, as shown by the shriveled neck, and the absence of a well defined protoplasmic mass in the basal cavity.

[8] **Their number can be** ascertained by the method used for Spirogyra, p. 36.

5. Section or crush a *young archegonial head* not exceeding a pinhead in size, and giving attention only to the archegonia, notice
   a. **The** single layer of cells forming the wall of **the** *bulbous part*, passing into
   b. **The** few rows of cells forming the *neck*, appearing in optical section like two rows, ending above in
   c. **The** stigmatic cells, which are spread apart at the time of fertilization.
   d. The well defined *cavity* in the bulbous part, containing (if not yet fertilized)
   e. The globular oosphere.
   f. **The narrow** canal extending the length of the neck, through which the antherozoids reach the oosphere to fertilize it.

ANNOTATIONS.

In a morphological point **of view** Marchantia is a plant of unusual interest, on account of its remarkable **degree of** differentiation. Taking first the vegetative **portion,** we have in the thallus **a** structure that is typically shown in lichens and other plants belonging to the thallophytes. More strictly speaking the Marchantia **stem is** only thalloid, for there are the rudiments **of** leaves on **its** under side, while a true thallus **has** no leaves. The prostrate position of the stem has necessitated the specialization of the upper surface **for** the purposes of assimilation and respiration, and the **lower** surface for the absorption of moisture and the other nourishment **which** comes **with it.**

The chlorophyll bodies, like those of all higher plants,

consist of rounded grains of protoplasm in which the chlorophyll proper is contained, the protoplasmic body being readily seen after the pigment has been extracted by alcohol. Such grains are scattered throughout the thallus, but are only effectively developed in special cells, which arise from the floor of cavities formed by depressions in the surface of the thallus, and which are overarched by the epidermis at a very early stage of growth.[9] Communication with the outside air is secured by means of peculiar and highly developed stomata.[10] They are wider in the middle than at the upper and lower openings, each stoma forming a small air-chamber. The border to the outer opening is sharp edged and immobile, while the inner one is formed of inflated cells which act as regulators to the passage of air and moisture. Altogether a very perfect arrangement is thus made for the aëration of the chlorophyll tissue without undue loss of moisture.

The under surface of the stem is provided with copious hairs, those of the wings developed to give support,[11] toward which the internal thickenings and spiral constriction of the walls contribute, while those of the midrib, larger, smooth-walled, and somewhat colored, serve to fix the plant to the earth and to take up from it the water and nutriment required, i. e. to perform the office of roots. In a physiological point of view

---

[9] Leitgeb, Die Athemöffnungen der Marchantiaceen, in Sitzber. d. k. k. Akad. in Wien, lxxxi, 1880. This differs from the older view which ascribed the openings to a separation of the epidermis from the underlying tissues. Sachs, Text-book 1st and 2nd Eng. eds.

[10] Described and illustrated by Voigt, Beitrag zur vergleichenden Anatomie der Marchantiaceen in Bot. Zeit., 1879, p. 729.

[11] According to Strasburger, Das botanische Practicum, p. 314.

the root-hairs are not merely rhizoids but real roots, and such they have been called by Sachs recently, who no longer holds to the morphological distinctions of root, stem, leaf and hairs, but refers all vegetative organs of higher plants to two categories, viz : the *root* and the *shoot.*[12]

The scales are organs that we shall meet with in a more developed form when we reach the ferns. They differ from the leaves in size and position, but more especially in having the cells empty and lifeless.

The internal structure of the stem is interesting on account of the thickenings of the cell-walls for securing extra strength, and the absence of any differentiation of the tissues along the midrib except the moderate change in the shape of the cells.

The branching of the stem is a fine example of true dichotomy where the growing point is symmetrically halved, and each half gives rise to a branch. [13] In this case one branch develops faster than the other, and the appearance is soon the same as if it had arisen as a lateral branch (see fig. 2). The tissues of the wings reach their growth more rapidly than those of the midrib, and so the growing end is constantly indented.

The leaves have little of the appearance we associate with the term, as commonly used. They are, indeed, very depauperate leaves, and serve simply as organs of strength, through the power of the protoplasmic contents of the cells to maintain turgidity.

The asexual propagation in Marchantia is of two

---

[12] Vorlesungen über Pflanzenphysiologie. p. 5.

[13] Sachs, Text-book, 2nd Eng. ed., pp. 177, 181.

kinds. One is a very common method, by which the
stems die off at the older end as fast as they grow at
the other. In this way the branches are eventually
separated from each other and become independent
plants. The other is a peculiar method by which certain hairs at the bottom of cupules grow into flat green
plates, the gemmæ, which as they become mature are
pushed out of the cupules by the aid of the secretion
from the glandular hairs. [14] The gemmæ have their
direction of growth changed at a very early stage by
the formation of a right and left growing point, so that
the young plantlet is bifurcated at its outset. When
a gemma has fallen upon the ground, the side which
happens to be uppermost is developed as the upper
surface of the thallus, and the other becomes the lower
surface. [15] The root-hairs grow from the cells devoid
of chlorophyll.

The sexual reproduction is among the most highly
developed of that shown by the liverworts. The organs
are upon branches whose modification is so interesting
that it will be necessary to examine it somewhat carefully. The plants are diœcious, bearing the reproductive
organs on separate individuals. In each case the reproductive branch consists essentially of an attenuated portion, the pedicel, terminated by an expanded portion, the
head, on which last the sexual organs are borne. The
pedicel is not a single branch, but two which are the
result of dichotomy at the point where it leaves the

---

[14] Fide Strasburger, Das botanische Practicum, p. 436.

[15] Engelmann, Ueber die Einwirkung des Lichtes auf den March-
antienthallus in Arb. d. bot. Inst. in Würzburg, Bd. ii, p. 665 ; Mirbel,
Mém. Acad. Sci. de Fr., xiii (1835), p. 355.

thallus. These two do not separate, and, indeed, were it not for the two double rows of leaves along the anterior (under) surface, which give rise to the two grooves with their strengthening hairs, it would be difficult to show that any branching had occurred.

The pedicel of the female head is made up of extensions of the tissues of the thallus, but without the development of the wings. The head is formed by sudden branching, and as dichotomous branching must always be in pairs, it results in an even number of branches which are spread out like a very widely open fan. But counting the rays of the head always gives an odd number, which is explained by the fact that the growing point is not at the tip of the rays but at the sinus between them, while the rays are formed, as in the thallus, by the extension of tissue on either side the growing point. Thus each ray, with the exception of the ones nearest the cleft of the head, stands between two growing points, while those next the cleft have a growing point only on one side of them. The hairs of the rays correspond with the hairs of the wings, and extend into the grooves of the pedicel.

If now we turn to the male branch, we shall find the pedicel only differs from that of the female in possessing no chlorophyll tissue on its posterior (upper) surface. The tissues of the upper surface of the head were at an early period of growth continuous with those of the thallus, but, owing to some unknown cause, they have not continued to expand along with those of the ventral side in forming the pedicel. The head is made up of branches, as in the female head, and like that is not a radial structure, but zygomorphic. The cleft is not so evident as in the other case, and the number of rays is even and not odd, the latter being the result of the growing point being at the ends of the rays, instead of at the sinuses. The various correlated changes can readily be worked out by the student.

It now remains to account for the position of the two kinds of organs, one being on the upper surface and the other on the lower. We must know in the first place that the antheridia are modified hairs, which originally started on the surface, but became inclosed in cavities by the surrounding tissues growing up about them. They evidently belong to the upper surface from their position, and the fact that those nearest the growing edge are the youngest. In the female inflorescence we find that the organs nearest the edge are not the youngest, but the oldest. We can only explain this by supposing that they belong to the upper surface, but are brought below by the turning under of the growing point.[16] The perichætium is the thin expanded edge of the thallus.

The antheridia and archegonia originate, as in the case of the gemmæ, from papilliform hairs, which divide into two cells by a transverse wall, the lower cell becoming the pedicel, and the upper the body of the organ.[17] Paraphyses, which are always sterile bodies, are very common among the cryptogams; their significance is not understood.

The antherozoids may be taken as the type of the motile male element in fertilization. They are formed

---

[16] Strasburger, Das botanische Practicum, p. 439; Leitgeb, Untersuchungen über die Lebermoose, vi, 1881.

[17] Sachs, Text-book, 2nd Eng. ed., p. 348.

of free protoplasm, having no cellulose covering. The hyaline vesicle which is sometimes seen attached to them arises from the internal part of the protoplasm of the cell, the outer portion of which produces the cilia, and the nucleus at the center of the cell the body of the antherozoid.[18]

The archegonia separate a mass of protoplasm in their interior, the oosphere, which is essentially a naked cell. After fertilization it divides in a perfectly regular manner to form the sporogonium. The fertilization is prepared for by the conversion of the axial row of cells of the neck into mucilage, the swelling of which forces the stigmatic cells apart, and a passage-way is formed to the naked oosphere. The antherozoids pass through this channel, become buried in the oosphere, and the fertilization is complete.

The elaters by their strongly hygroscopic character assist materially in forcing out and distributing the spores.[19]

---

[18] Strasburger, Das botanische Practicum, p. 455.

[19] The student should consult Hofmeister's Higher Cryptogamia, which contains a very full statement of the development of Marchantia, with historical references up to 1862.

MOSS.

Atrichum undulatum Beauv.

PRELIMINARY.

MOSSES appear so much alike to those who have not given special attention to them, that it is more difficult to definitely point out a particular species than in the other plants of the book. The one selected for study is widely distributed, and very common, forming carpet-like patches in woods, and on shady banks. The single plants stand from two and a half to four centimeters (one to one and a half inches) high. The leaves, which are abundant, are five millimeters (quarter of an inch) or more long, narrow, with wavy sides; the undulations appear, when the leaf is held to the light, as lines passing obliquely from the middle to the margin.

The male and female plants are usually found in separate patches, as in Marchantia. The male flowers (see fig. 5) are easily recognized by being cup shape, and are distinguished from the rosette of leaves terminating a rapidly growing stem by having a distinct, rather flat bottom to the cup. They are readily found at almost any time during the year, and are especially abundant in early summer.

The female flowers, which are less common than the male, differ so little in external appearance from the ordinary vegetative condition, that it often requires a protracted search to find them. A patch of female plants may usually be detected by the presence of the fruit in some condition of growth or decay ; if, on cutting vertically through a stem taken from such a group of plants, the terminal leaves of which are well folded over the end, making a loose bud, the stem appears to terminate abruptly within the bud, it may be inferred that the female flowers are found. It is, however, necessary to use the microscope to render it sure. They are to be sought for especially in May. If the female flowers can not be found, those of other mosses will answer the purpose. *Polytrichum* is one of the largest of our mosses, and has female flowers much like Atrichum, while *Mnium*, *Funaria*, and others have them somewhat larger, more conspicuous, and nearly as common as the male.

The fruit is a nearly straight cylindrical pod with a conspicuous pointed beak, borne erect on a stalk about two or three centimeters (an inch) long (see fig. 4). Collect both green fruit from which the hood (calyptra) has not fallen, and that which is thoroughly ripe with the hood and lid both gone, exposing the teeth.

The protonema is not so abundantly produced as in many mosses. Keeping vigorous growing plants in an inverted position in a moist atmosphere for some time by turning a bell-glass over them, will sometimes be sufficient to develop it. The protonema from other mosses (*Mnium*, *Barbula*, etc.) is, however, usually found with ease, or may be produced as above, and will serve for the study.

The materials required for the present study are

alcoholic specimens showing male and female **flowers,**
and fruit ; fresh specimens showing protonema and male
flowers ; potassic hydrate ; and iodine.

LABORATORY WORK.

GROSS ANATOMY.

A. GENERAL CHARACTERS. Note

1. The vertical *stem ;* unbranched, or branching only from
the base.

2. The *leaves* clothing the stem.

3. The root-hairs, *rhizoids ;* often forming a close felt at
the base of the stem.

4. At the summit of some of the stems, the *flowering heads*
of two sorts :
   a. The male heads forming a terminal rosette of green
   leaves.
   b. The female heads with the terminal leaves folded
   over each other forming a small bud.

5. At the summit of other stems the fruit or *sporogonium,*
consisting of
   a. The slender stalk or seta.
   b. The cylindrical pod or capsule.
   c. The hood which fits closely over the end of the
   pod, and may be easily pulled off, or has dropped
   off of itself, the calyptra.

6. Among thrifty plants that have been kept under a
moist bell jar for several days, notice the green threads
growing out over the soil, the protonema.

B. THE RHIZOIDS. Remove some from the stem,
mount, and notice the small tangled hairs forming the mass.

C. THE STEM. Notice •

1. The *size* and *shape*.

Remove the leaves near the base, mount **a transverse section,** and notice

2. The *outline* of the section.

3. The three *tissue* **regions** ; the peripheral brown tissue, **the** axial tissue forming a light spot in the center, and the intermediate colorless tissue.

D. THE LEAVES. **Notice**

1. The manner of *arrangement* on the stem.

2. The difference in *size* on different parts of the stem.

3. The **shape** of
   a. The lowest, **scale leaves**.
   b. The middle, **foliage leaves**.
   c. The uppermost on flowering stems, forming **the** outer portion of **the** head, **perichætial leaves**.

4. The structure ; a thin *lamina*, with a thicker **median** line, the *midrib*.

5. The character of the *margin*, especially toward the apex.

6. In the foliage leaves, the *undulations* passing obliquely **outward from the midrib to the margin ;** their absence **in the other sorts.**

7. Draw a leaf of each sort—scale, foliage and perichætial.

E. THE FLOWERING HEAD.[1]

1. **The** *male heads*. **Notice**

---

[1] Called the " receptacle " by Sachs (Text-book, 2nd Eng. ed., p.370), but this term has long been in use for the end of the stem on which the parts of a flower are seated. The analogy of the several parts of the moss " flower " to those of the head of a composite (e. g. sunflower) has determined the use of corresponding terms.

a.   The *shape*.

b.   The *central disk*.

c.   The *leafy continuation* of the stem arising from the center of some of the heads.

Cut a head in two vertically, and note

d.   The enlarged end of the stem, *receptacle*, on which the disk is seated.

e.   Draw the half head, looking at the cut surface.

Remove the disk with the point of a scalpel, separate the parts on a slide, mount, and notice

f.   The broad chaff, resembling the scale leaves ; the shape, especially the narrowed base.   Draw.

g.   Numerous narrow bodies of nearly the same length as the chaff, *antheridia*, the male reproductive bodies.

h.   Slender filaments of same length as the antheridia, *paraphyses*.

2.   The *female heads*.   Make a vertical cut exactly through the center, and notice

a.   The absence of any thickening of the stem to form an enlarged receptacle.

b.   The absence of a disk.

Remove the central portion, separate well on a slide, mount, and notice

c.   The numerous filaments, the *paraphyses*.

d.   A few bodies, not exceeding a half dozen,[2] about as large as the antheridia, but swollen somewhat near the base with the upper portion slender, the *archegonia*, the female reproductive bodies.

F. THE FRUIT.   Notice

---

[2] The fewness of the archegonia, and the difficulty of securing them at just the right stage of growth, often makes an extended search necessary in order to demonstrate them.

1. The stalk or *seta*.
   a. The *length.*
   b. Character of the **surface**.
   c. The slightly expanded **end from** which **the capsule** arises, **apophysis**.

Take a specimen that has been boiled a minute or two in potassic hydrate, and pull the seta from the leafy portion, taking care that it does not break off, but comes **away** smooth, and notice

   d. The pointed *base.*

2. The pod carried by the **seta**, the *capsule*, with its calyptra ; notice
   a. The manner in which the *calyptra* fits upon **the** apex **of the** capsule.
   b. *Shape* of the calyptra.
   c. Pull away a calyptra and note its *texture*, and **the** roughness **of its** *apex.* Draw.
   d. The shape **of** the capsule, and nature **of the** surface.
   e. The hemispherical apex bearing a **long** beak, together forming a removable lid, the **operculum.**
   f. The obliquity and slightly eccentric position of the *beak.*
   g. Draw a capsule.
   h. Pull off the operculum from **a mature fruit, and** notice the *rim* of the capsule on which **the edge** of it rested.
   i. Rising from the rim, a large number of delicate, incurved **teeth,** together forming the **peristome.**
   j. Count the *teeth ;* the *number* **will be** some multiple **of four.**
   k. The delicate **epiphragm** stretched **between the** apices of the teeth ; to be better displayed shortly.

l.   Draw the upper part of the capsule showing the teeth and epiphragm.

Divide the capsule longitudinally, and notice

m.   The axial column running through the center, the **columella**, the expanded apex giving rise to the epiphragm.

n.   The cavity between the columella and wall of the capsule, either empty or filled with a powder, the *spores.*

o.   Make a diagram of the section.

MINUTE ANATOMY.

A. THE RHIZOIDS.   Under high power, notice

1.   The straightness, uniformity of diameter, and mode of branching.

2.   The character of the lateral *walls*, and the position and direction of the cross-partitions, if any.

3.   Draw.

4.   Notice the manner in which some of the rhizoids are coiled around each other, forming ropes.

B. THE PROTONEMA. Notice

1.   The arrangement of the cells.

2.   The thinness of the walls, and position of the cross-partitions.

3.   The contents.

4.   Draw.

C. THE STEM. In a transverse section taken from the lower scaly part of the stem, notice

1.   The *three regions.*

a.   The *peripheral*, with the cell walls reddened.

b. The *axial*, with the cell walls colorless.

c. The *intermediate*, with cell walls yellowish.

2. The *peripheral tissue.* Note

   a. The outer layer, **epidermis**, occasionally **bearing** root-hairs.

   b. The similar underlying cells, merging **into**

3. The *intermediate region.* Note

   a. The larger cells with the walls becoming thinner **toward** the center of the stem.

   b. One or more **leaf traces**, composed of

      i. A crescent shaped layer of small round cells **with very** thick walls, the **dorsal cells.**

      ii. Lying in the crescent, about two rows of larger cells with rather thin walls, the **conducting cells**, inclosing

      iii. Two **or** three small cells, appearing **much** like intercellular spaces, the **central cells.**

      iv. Still further toward the center of the stem, **a** few scattered cells similar to the dorsal, the **basal cells.**

   c. Note that the leaf traces nearest the center of the stem are the simplest. Sometimes one may be found at the very center of the stem.

4. **Draw one of** the largest leaf traces with some of the surrounding tissue including the adjacent epidermis.

5. The *axial region.* **Note**

   a. The more or less strongly thickened walls **of the** cells.

   b. The small groups of cells with the intervening walls very thin and membranous.

   c. Draw a portion of the axial region.

6. In a longitudinal section of the stem, identify as **many of** the different sorts of cells as possible, noticing

    a.   The shape of the cells.   Draw.

    b.   The direction taken by the leaf traces.

Remove the scales from a stem, cut a slice from the surface, and notice

  7.   The shape of the *epidermal cells.* **Draw.**

D. THE LEAF. Make a transverse section just below the middle of one of the largest foliage leaves, and notice

  1. **The** larger central portion, **the** *midrib.*

  2. **The** plate of cells, usually a single row, extending right and left from the midrib, the *lamina.*

  3. The *midrib.* **Note.**

    a. **The** *epidermis :* a single layer of cells on the convex (under) side ; a layer on the flat (upper) side, each cell of which gives rise to a vertical plate, two to four or more cells in height. If from a fresh specimen, note the contents of the cells.

    b. The *leaf bundle;* compare the several parts, the dorsal, basal, central and conducting cells, with the corresponding parts of the leaf trace which enters the stem, already examined.

    c. Occasionally a few cells between the dorsal and basal cells and the adjacent epidermis, resembling the latter.

    d. Draw the midrib.

  4. The *lamina.* Note

    a. The shape and contents of the cells.

    b. The smaller grouped cells at the margin of the lamina.

    c. Draw.

Mount a *foliage leaf* entire with the upper side uppermost, and beside it another with the lower side uppermost ; using low power, notice

 d. **The cells of the *main part*** of the lamina with their
contents.

 e. **The** *marginal cells*, produced into

 f. The sharp forward-pointing *teeth*, which are often
in pairs . observe the distribution of the teeth
along the margin, also similar teeth on the under
side of the leaf along the summits of the undulations and on **a part** of the midrib.

 g. The surface of **the *midrib*** : observe **the** shape of
the cells on the under side ; the rows of chlorophyll
tissue on the upper side, which begin near the base
of the leaf and extend nearly to the apex, seen as
plates in the transverse section.

Under high power, notice

 h. The elongated marginal cells, and the shorter cells
forming the teeth.

 i. Draw a portion from near **the middle of the leaf,**
showing teeth, marginal cells and some **adjacent**
laminal cells.

E. THE FLOWERING HEAD. Remove the disk from
a male head, and mount, well separated ; notice

 1. Numerous hairs, the *paraphyses.*

  a. The walls, cross partitions, and contents.

  b. **Draw.**

 2. The *antheridium.*

  a. The *shape.*

  b. The elongated cells of the *body.*

  c. The short cells of the *pedicel.*

  d. **The large *apical*** *cell,* in antheridia which have not
yet burst.

  e. **Draw.**

 3. The *antherozoids.* **If from a** fresh specimen, notice

    a.   The *movement*. Apply iodine, and watch them as they gradually come under its influence.

    b.   The *form ;* a slender body, with a pair of cilia at the anterior end.

    c.   The colorless *vesicle* sometimes to be seen attached to the posterior part.

    d.   Draw.

Crush an immature antheridium by pressing on the cover-glass, and as the contents escape, notice

    e.   The antherozoids coiled within the mother-cell.

If alcoholic specimens are used, the antherozoids may be seen within the mother-cell, but the parts can not be made out.

Tear apart a *female head*, mount, and notice

4. The *paraphyses ;* shape and structure.

5. The *archegonia.*

    a.   The enlarged *ventral portion.*

    b.   The long *neck.*

    c.   The short thick *pedicel.*

    d.   Focus upon the surface, and draw some cells of each portion.

Treat with potassic hydrate to render the archegonia more transparent, focus so as to give an optical section, and notice

    e.   The two rows of cells forming the *neck*, the terminal cells of which are

    f.   The *stigmatic cells.*

    g.   The *canal* along the axis of the neck.

    h.   The two or more rows of cells surrounding the *ventral portion.*

    i.   The small mass of protoplasm lying deep in the center of the ventral portion, the *oosphere* (if not yet fertilized).

j. Draw.

F. THE FRUIT.

1. The *seta.* In a transverse section near the base, notice

    a. The outer portion of thick **walled, deeply** colored tissue, passing abruptly into

    b. **Loose, thin** walled, colorless tissue. Within these and almost completely separated from them

    c. A core composed of the following tissues :

        i. An outer row of large, round, thin walled cells.

        ii. Adjoining this a layer of smaller angular cells with walls somewhat thickened, and

        iii. In **the** center, **a few** small cells with thin **colorless** walls.

    d. **Draw a** sector of the section.

    c. Notice **the shape of** the **epidermal cells in a** surface slice. Draw.

    f. Examine several longitudinal sections, **and determine as many of** the tissues **as possible. Draw.**

2. The *capsule.* Make a transverse section through the middle of an immature capsule. Under low power, notice

    a. Two parts, separated by a cavity :

        i. The **outer, the** *wall of the* **capsule.**

        ii. The inner, **the** *axial cylinder.*

        iii. Uniting these, if not torn away in making the section, delicate *radial filaments.*

    b. **The parts** of the axial cylinder.

        i. **The** narrow outer part, the *wall of the spore case.*

        ii. **The** large central part, the *columella.*

        iii. **A** dark line **separating the** two, the *mother-cells* containing the young spores.

c.   Make a diagram of the section.

Under high power, examine in succession

     d.   Each of the tissues enumerated.

     e.   Draw a sector of the section.

Make a transverse section of a *mature capsule*, notice

     f.   The thick walled, deep colored and strongly cuti-cularized epidermis.

     g.   The colored cells of the spore sac.

Make a longitudinal section of a *nearly mature capsule* (after removing the calyptra), and with low power, notice at the base of the capsule

     h.   A mass of large thin walled cells forming the *apophysis.*

     i.   Above the apophysis several layers of smaller, more regular cells, from which arise the various parts of the axial cylinder.

     j.   At the upper end of the capsule, notice

         i.   The large central mass of wide thin walled cells, resting upon the axial cylinder and inclosed by the *operculum.*

         ii.   The line of separation between this and the roof of the operculum, showing, more or less clearly, the delicate membrane which is exposed by the detachment of the operculum, the *epiphragm.*

         iii. The small deeply colored cells of the *rim* of the capsule.

         iv. The curved lines extending from the rim to the edge of the epiphragm, the structure usually not well shown, the *teeth of the peri-stome.*

         v.   The tissue of the operculum on the sides where it shuts over the teeth, of the roof adjoining the epiphragm, and of the beak.

k. Illustrate the arrangement of the tissues as seen in longitudinal section by a diagram.

Take a nearly mature capsule, remove the thinnest possible slice from the side of the operculum with the razor inclined toward the beak ; the next slice will include a portion of the peristome, in which notice

l. The rows of cells from which the teeth are formed, and their manner of thickening. Draw.

Make several transverse sections through the rim and operculum, and study

m. The formation of the teeth from groups of cells.

Take a mature capsule, mount a number of entire teeth, and notice

n. The shape and structure of the teeth. Draw.

o. Flatten out a calyptra, and observe the cellular structure, especially at the apex. Draw some of the cells.

3. The *mature spores ;* note under high power
   a. The *shape.*
   b. The *wall* and *contents.*

ANNOTATIONS.

The step from Marchantia to Atrichum is not so great as that which intervenes between the several preceding examples, and yet the advancement is well marked and especially significant. With the upright growth of Atrichum is correlated the disposition of the leaves and root-hairs. The leaves being green, relieves the stem of its assimilative duties, and in consequence the smaller size and greater firmness better meet the requirements. The root-hairs simulate true roots even more closely than those of Marchantia. A curious habit of the root-hairs of this and the allied genera is the manner in which they coil about each other, forming branching ropes, and adding to their effectiveness as hold-fasts.

The stem of Atrichum shows considerable diversity of tissues. The axial groups of cells with thin intermediate walls are peculiar to a few of the higher mosses. A noticeable feature is the absence of a well marked epidermis, which is doubtless to be associated with the fact that the cells beneath have thick walls, that there are no chlorophyll tissues to be aërated, and that the numerous leaves assist materially in giving protection. The absence of stomata is also to be accounted for by the absence of chlorophyll tissues.

The leaves show a distinct midrib and blade, and possess all the essential features of true foliage leaves. The blade being only one cell thick is apparently the same on both sides, and possesses chlorophyll bodies which are typical for all higher plants. A selvage of strong cells runs around the edge of the lamina to guard against tearing, while numerous teeth act, to some extent, as a protection. To give additional aërating surface, there are a number of plates, like narrow auxiliary blades, placed along the upper surface of the midrib. They are still better developed in *Polytrichum*, but are entirely wanting in most mosses. As there is no epidermis or other protective structure to guard against excessive evaporation, an ingenious substitute is afforded by the inrolling of the sides of the leaf whenever the turgidity of the cells is disturbed.

But no feature in the histology of mosses is more

significant and interesting than the leaf bundle of the midrib. It is the simplest form of a structure that plays a most important part in higher plants—the framework of wood and bark which enables them to rise above the surface of the earth and display their tissues to the wind and sun under conditions most favorable for growth. The bundles of Atrichum which are as highly developed as in any of the mosses, resemble those of higher plants more in their position and function than in structure.[3] Their place in the leaf and their manner of forming leaf-traces in the stem are like those of higher plants. The cells for strength are the dorsal and ventral, being the same except in position, and the cells inclosed by these transport the sap.

Passing to the sexual reproduction, we notice that the organs concerned are much like those of Marchantia. The differences requiring consideration lie in the modes of displaying and protecting the organs. Instead of sinking the male organs in a flattened receptacle, they are placed in the axils of protecting leaves diverted to that use, and instead of bringing the female organs under the protecting roof of the receptacle they are sheltered from rain and other excessive moisture by the overlapping of the perichætial leaves.

An item of historical interest in this connection is that it was in the mosses that the sexual organs of cryptogams were first demonstrated by Hedwig[4] in 1783, but it was not till the publication of Suminski's researches on the ferns,[5] as late as 1848, that their

---

[3] A very full illustrated account of the histology of the stem and leaves of mosses is given in Pringsheim's Jahrb. f. wis. Bot., vi.

[4] Theoria Generationis, p. 138.

[5] Zur Entw. der Farrnkräuter.

sexual character was fully established. It was also in mosses and liverworts that the antherozoids were first detected, being seen by Schmidel[6] in 1762, but without detecting their cilia, which were discovered by Unger[7] in 1837.

After fertilization has occurred the oosphere clothes itself with a cell wall, and grows at once into a fruit, as in Marchantia. This fruit is in many ways remarkable, as will be more apparent in some respects after studying the ferns and club-mosses. It will be remembered that in the plants already studied, with the exception of Marchantia, the sexually formed spore produced a plant like the parent, after a longer or shorter period of rest. In Atrichum, however, it grows, not into a plant like the parent, but into a highly complicated structure, the fruit or sporogonium, which in its turn forms asexual spores that produce plants like the original. This process, known as an alternation of generations,[8] is less strongly marked in liverworts, and reaches its height in ferns.

The base of the seta which is thrust into the apex of the leafy plant, has no organic connection with it, and while in Atrichum it pulls out with some difficulty, in many mosses it comes away easily without preparatory treatment. This feature further emphasizes the distinctness of the so-called fruit and the parent plant, from which in quite a parasitic fashion it derives its nourishment.

---

[6] Icones plantarum, p. 85.

[7] Nova Acta A. C. L.–C. Nat. Cur., xviii, p. 791.

[8] Sachs, Text-book, 2nd Eng. ed., pp. 226, 954; Vines, Journal of Botany, 1879 ; Underwood, Our native ferns and their allies, p. 35.

The tissues of the seta attain rather higher **development** than those of the stem. The cortical part **is** provided with a well formed epidermis, while the axial part is composed of several tissues, the two portions being separated by thin-walled parenchyma. At the apophysis, where the seta expands at its upper end, **many** mosses produce stomata quite like those of higher plants. In rarer instances they occur on **the** capsule **or** seta. Their presence or absence seems to signify nothing as to relationship, as **there** is no more constancy in their occurrence among the highest than among the lowest forms.[9]

The capsule of Atrichum does not differ widely from that of other mosses, except in the teeth **and** epiphragm, and otherwise **requires no particular** explanation. The **teeth are composed of groups of** cells arranged as a series of **U**'s placed side **by** side. In all other mosses except the immediate allies, **where teeth** are present at all, they **are formed of** the thickened **sides of** the cells, and not **of** whole cells.[10] The epiphragm, which joins the apices **of** the teeth like a thin membrane, is formed without thickening or special preparation of the walls. The spores escape by being shaken from the capsule through the openings between **the teeth, as from a pepper box.**

The calyptra, which **is the** result **of the** aftergrowth **of the** archegonium, **was** early **torn away from its** attachment at the base **of the** fruit **and** carried up by the elongating **seta as a hood** for the capsule.

---

[9] Valentine, Trans. Linn. Soc., **xviii**, p. 239.
[10] **Sachs**, Text-book, 2nd Eng **ed., p.** 383.

The spores germinate by producing a protonema, which may grow to considerable length, with numerous branches, before a leafy stem is formed. The successive inclinations of the transverse walls of the protonema have been shown to follow the same laws as govern the successive divisions of the apical cell to form the leafy stems, so that we are to consider the protonema as an excessively attenuated stem, from which the leafy stems arise as lateral branches.

THE MAIDEN-HAIR FERN.

Adiantum pedatum I..

PRELIMINARY.

THE maiden-hair fern is abundant in dark rich woods throughout the eastern part of the United States, and occurs to a considerable extent west of the Rocky Mountains. It may be recognized with certainty by the forking of the polished purple leaf-stalk into two equal recurved branches, which give rise to a number of straight branches upon one side, bearing the oblong leaflets. On the back of the leaflets, along their margins, are born the crescent-shaped fruit dots.

Underground stems and roots (together popularly called roots), and leaves, including the leaf-stalks, should be collected when the fruit dots assume a yellowish brown hue, which is usually about the middle or latter part of August. The roots should be taken up with care and the dirt shaken from them gently to avoid tearing off the root-hairs and root tips, and the cleaning completed with water. Part of the leaves and all of the stems and roots should be preserved in alcohol, the remainder of the leaves by drying between newspapers or in a plant press.

The prothallia of Adiantum are less known popularly. They are flat, roundish, green bodies, two to five millimeters ($1/16$ to $1/5$ inch) in diameter, deeply notched on one edge, and held to the ground by a cluster of root-hairs from the under side. They may be found on the surface of damp ground near patches of the fern, and may be collected and preserved in alcohol. If a green-house is accessible, prothallia may usually be obtained fresh and in quantity from the surface of pots and earth near which native or exotic species of Adiantum are growing. If neither source yields suitable material, the prothallia may be grown by sowing the spores of Adiantum (to be obtained from the fruit dots on the margins of the leaflets) on the surface of damp earth packed smooth and kept at first under a bell-glass in a good light.[1] Strasburger[2] recommends sowing the spores on the surface of a piece of pressed peat (previously boiled in water to destroy other spores) which is to be kept saturated with a nutritive solution prepared according to the formula given on page 34. The peat should be covered by a bell-glass and placed near a north window. If prothallia of Adiantum can not be obtained, the prothallia of almost any fern will show the characteristic features of this stage.

It will be advisable before attempting to cut sections of the rhizome to soak it for a few minutes in water in order to soften the tissues somewhat, for when taken from alcohol they are extremely hard. Care will have to be exercised in cutting these sections not to nick the edge of the razor; it will need frequent sharpening. Before cutting the sections, the end from which they are to be cut should be smoothed with a knife.

[1] Cf. Campbell, Bot. Gazette, x, p. 356.
[2] Das botanische Practicum, p. 457.

The requisites for the complete study of this plant are dried and alcoholic specimens of **leaves**; alcoholic specimens of roots and stems; fresh prothallia; alcohol; iodine; potassic hydrate; and solution of potassic chlorate.

LABORATORY WORK.

GROSS ANATOMY.

A. GENERAL CHARACTERS. Taking a complete plant, notice the four parts into which it may be readily divided :

1. The horizontal, very dark brown, or almost black, under-ground stem, the **rhizome,** from which are given off

2. A number of slender branching fibers, the **roots**.

3. The aerial portion, **the leaf or frond,** consisting **of** slender polished *stalks*, and flat **green** expansions, the *blades.*

4. The appendages to the surface, *trichomes*, in the form of scales on the rhizome, hairs on the roots, and reproductive bodies on the leaves.

B. **THE STEM** or RHIZOME. Notice

1. The *size, shape* and *surface.*

2. The occasional *branching.*

3. The **nodes** and **internodes**; the nodes **are** indicated **by the** growth of a leaf at each, alternately **on the right** and left sides ; the intervals between the nodes **are the** internodes.

4. The growing *apex ;* **the** dying *base.*

5. The *buds* near and at the apex. **Strip** off carefully

from several buds the numerous brown scales which clothe them.   Note the two kinds :

a.   Buds showing a *rudimentary leaf* whose stalk is coiled upon itself, thus : ♉

b.   One or more buds whose central part is simply a continuation of the stem.

6. Make an outline drawing of the rhizome, showing the size, shape, mode of branching and arrangement of leaves and buds.

7. The *structure*.   Cut across the rhizome at right angles to its length and examine the cut surface   Observe

a.   The outer ring of brown tissue, the *cortical layer*.

b.   The oval, circular, or **C** -shaped white mass, the fibro-vascular bundle.   Where a branch or leaf arises two fibro-vascular bundles will be seen, thus : **C Ɔ**.   Find a part of the rhizome showing such an arrangement, and trace the course of the bundles (by cutting a series of rather thick sections) through at least two internodes, noting the modes in which successive branches are given off from the bundle. The smaller **C**-shaped portion passes into the nearest leaf ; the other gradually enlarges, closes into a circle, elongates into an oval, becomes egg-shape, and finally opens to form two unequal **C**'s, the smaller of which soon enters the second leaf on the opposite side of the rhizome from the first.

c.   Inclosed by the fibro-vascular bundle a darker brown mass not differing otherwise from that surrounding the bundle.

d.   Make an enlarged drawing of the cut end of the stem.

Cut a rhizome longitudinally through the center, and on the cut surface make out

e. The parts previously seen in the transverse **section**. Draw.

f. The *scales*. Mount a **few** scales from the rhizome, and note

    i. Their *shape* and *texture*.

    ii. Their *structure ;* the shape and arrangement of the cells.

    iii. Draw a scale enlarged.

C. **THE** ROOTS. Notice

1. The *shape*.

2. The mode of *branching*.

3. Their *position* on the rhizome.

4. **The covering** of tangled *root-hairs* **with which some are** enveloped.

5. The absence of root-hairs **near** the whitish *growing* ***end***.

6. **The** brownish tip (sometimes torn off), **the root-cap.**

7. Examine a tranverse section of a root, and compare with that of the stem. Note the position of the fibrovascular bundle. Draw the section.

D. **THE LEAF.** It may be easily distinguished into two parts, the stalk, **rhachis**, with its branches, and the green blades, **pinnules.**

1. **The main *rhachis* and its branches. Note**

    a. The polished *surface*.

    b. **The** *shape ;* **a** little flattened on the anterior **surface (i.e., the one** corresponding **to the** upper surface **of the leaf).** Dried or alcoholic specimens are likely to have this surface flat or concave while **the** posterior **remains** convex. **Note the** slight *ridges* between these two surfaces, more marked in dried or alcoholic than in fresh specimens.

c.   The *color* of the anterior and posterior surfaces.

d.   The *branching*.   At the top the rhachis divides into two equal (or almost equal) divergent branches.   Each of these again divides into two, one of which forms the rhachis of a pinna (to be described shortly), while the other again forks. Note the number of times such forking occurs and the relative length of the secondary rhachises thus formed.   Make a diagram showing the above points.

e.   The *structure*.   Cut transverse sections of the stalk at various heights.   Make out the same structure as detailed for the rhizome.   Notice

  i.   That the brown tissue of the stem is largely replaced by a whitish one, *parenchyma*.

  ii.   The different shape of the sections of the *fibro-vascular bundle* at various heights along the stalk.

  iii.   Trace its *course* near the forking of the stalk until it divides, one-half entering each branch.

  iv.   Make diagrams showing these points.

f.   Compare the *scales* on the base of the leaf-stalk with those studied from the rhizome.

2. The *pinnæ*.   Each pinna is composed of a slender polished rhachis bearing a number of leaflets, the *pinnules*. Note the variation in the number of pinnules on a rhachis and the general outline of a pinna.   Make an outline drawing of a pinna.

3. The *pinnules*.   Selecting a pinnule near the middle of the rhachis, observe

a. The *shape* as to outline and margin.

b. **Draw** carefully an outline of the pinnule studied.

c. Compare the shape of the terminal pinnule **with** those near the middle of the rhachis. Note that it is like two of the latter joined by their bases. Compare also the basal pinnules with the middle ones. **Draw an** outline of the terminal **and** basal pinnules.

d. The *surface, texture,* **and** *color.*

e. The *structure.* Notice

   i. The slender *stalk* at the angle formed by the lower and basal edges, attaching the pinnule to the rhachis.

   ii. The slender branching threads, **veins,** extending from the apex of this stalk **and** supporting

   iii. The green substance of the pinnule, **the mesophyll,** which **fills all the space between the** veins.

f. The arrangement of the veins, **venation.** Notice

   i. One vein a little stronger than the rest, parallel with and close to the lower edge.

   ii. The mode of branching.

   iii. That the veinlets are not connected into **a** network.

   iv. Compare **with** the venation of the basal **and** terminal pinnules.

   v. In the outline drawings of the terminal, **basal** and middle pinnules already made draw the veins.

4. **The** *reproductive bodies.* Observe

a. On the upper edges of the under side of the pinnules a large **number** of crescent shaped spots,

sori.   Note the pinnules from which they are most
uniformly absent.

Soak a pinnule in water for a few minutes and with the
needles turn back

b.   The flap which covers a sorus, the indusium.   Notice
that it is a portion of the edge of the pinnule
reflexed and peculiarly modified.

c.   On the under side of the indusium, a mass of yellowish spheroidal bodies, the *sporangia.*

Scrape away most of the sporangia from the surface, and
notice

d.   The relation of the points of attachment of the
sporangia to the veins.   Cut off and draw an indusium showing this.

5.   The *sporangia.*   Mount some of the separated sporangia and examine by oblique light.   Note

a.   Their *shape.*

b.   The short *stalk* by which they were attached.

c.   The ridge, slightly darker than the rest, extending
part way round the sporangium, the annulus.

d.   Burst a sporangium and note the contents, minute
powdery bodies, the *spores.*

e.   Study the *manner of bursting* and scattering the
spores.   Tear a bit of an indusium from a dried
specimen previously soaked in water, retaining
only a few sporangia ; place it on a slip of glass
and allow it to dry, while watching the sporangia
through a lens, illuminating them from above.   A
crack appears on the side where the annulus is
absent, which gapes more and more as the annulus
straightens and becomes recurved.   After bending
backward a certain distance, by a sudden jerk
whereby the spores are scattered, the annulus

becomes straight **again** (or **almost so), and very** gradually resumes the same position as before **the** rupture of the sporangium.

E. **THE** PROTHALLIUM. Examine prothallia **of** various ages. Notice

1. The *shape* and *size.*

2. The cellular *structure*, **best seen** in **a** mounted specimen.

3. The cluster of *rhizoids* on the under side.

4. That in a prothallium **with** a *young fern* plant attached the plant arises from the under surface.

5. **Draw.**

MINUTE ANATOMY.

A. THE STEM. Cut **a** transverse **section and examine** with a low power. Make out **the** following parts :

1. The single outer row of cells, the *epidermis.*

2. A considerable thickness of brown[a] thick-walled tissue, the peripheral **sclerenchyma.**

3. A circular, oval or C-shaped mass **of** whitish tissue, most of which is the *fibro-vascular bundle.*

4. **Surrounding** this bundle, and marking its outline, **a** chain-like row of minute oval cells, the **bundle-sheath.**

5. Entirely **or** partially surrounded by the fibro-vascular bundle, a mass of axial sclerenchyma similar to the peripheral.

Examine the section with a high power and study in

---

[a] Yellowish in very thin sections.

detail each of the tissues and groups of tissues seen above, in the following order :

6. The *epidermis.*   Observe

    a.   That the outer wall is thicker than the lateral and inner ones.   In favorable sections a very thin layer, the **cuticle,** may be seen covering the outer wall.

    b.   That the epidermal cells contain numerous round-ish or somewhat angular *starch granules.*[4]   Treat a freshly-cut section with iodine, and notice the color produced.

        i.   Study one of the starch grains.   Notice the central lighter spot, the **nucleus.**[5]

    c.   Draw several epidermal cells.

7. The *sclerenchyma,* peripheral and axial.   **Note**

    a.   How greatly the *walls* are thickened.

    b.   That adjoining **walls** consist of three or more dis-tinct layers, the thin central one of which is the **middle lamella.**

    c.   The perforations or **pits,** which extend through the thickening layers to the middle lamella at right angles to the surface of the wall.   Observe that the pits in contiguous cell walls correspond to one another.

    d.   Examine the middle lamella at a point where three or four cells meet.   Note that it divides, inclosing a triangular or quadrangular space which is filled

---

[4] The occurrence of starch in epidermal cells is unusual.

[5] The term as here used has an entirely different signification from that which it has as applied to a cell.   Here it denotes a central watery spot, about which lie the layers of the starch grain, alternately more and less watery.

with a thickening deposit, similar to that of the inner layers of the wall.

e. The abundance of starch in this tissue.

f. Draw several sclerenchyma cells showing **these** points.

8. The *cortical parenchyma*, lying just outside **the** bundle-sheath in some **places.** Observe

    a. That the *walls* are thin and colorless, with trian-**gular** intercellular spaces; *contents* of the cells, granular protoplasm and starch.

    b. Compare carefully the middle lamella of the sclerenchyma with the walls of the parenchyma where the two tissues merge.

    c. Draw several parenchyma cells.

9. The *bundle-sheath.* Notice the **emptiness** of the cells, their shape, and the position of **their** longer axes.

10. The *fibro-vascular bundle ;* easily distinguishable **into** two regions : first, a central one, the **xylem,** characterized by the numerous large openings of the **scalariform vessels,** with small cells, the **xylem paren-chyma,** packed between them ; secondly, a peripheral one, the **phloem,** showing cells of much more uniform diameter, and lying between the xylem and the bundle-sheath. This region contains **phloem parenchyma** and **sieve cells.** Study carefully each **of** the above named tissues. Commencing at the bundle-sheath, examine

    a. The *phloem parenchyma ;* composed of two or three (occasionally but one) irregular rows of small thin-walled cells next the bundle-sheath and a few cells

---

[6] Sometimes a few starch grains appear to lie in them ; **they** have been pulled over by the razor in cutting.

here and there between the sieve cells (to be pointed out directly), all filled with granular pro- toplasm and small starch grains.   Compare with cortical parenchyma.

b.   The *sieve cells ;* lying between the main body of phloem parenchyma and the scalariform vessels. Note their angular shape, slightly thickened walls and emptiness, except for a little granular material clinging to the walls.

c.   The *scalariform vessels.*   Observe

   i.   That wherever two vessels are in contact their contiguous walls are flattened, and the vessels are therefore irregularly polygonal, having two or three sides much longer than the others.

   ii.   That they are thicker at the angles than on the sides, and thus appear to be united only at the angles.

   iii. The narrow slit between the contiguous sides of the vessels.

   iv.   The emptiness of the vessels.

d.   The *xylem parenchyma;* small cells packed between the scalariform vessels, and similar ones near their periphery.   Note their contents.

e.   Notice the general *arrangement* of the tissues, making it a concentric bundle.

f.   Draw sufficient of the fibro-vascular bundle and its sheath to show the different tissues and their relations to one another.

Cut a longitudinal radial section of the stem in the plane of the leaf stalks.   Examine with a low power, and make out

11. The epidermis.

12. The sclerenchyma, peripheral and **axial**.

13. The double band of whitish tissue, consisting of cortical parenchyma, bundle-sheath and fibro-vascular bundle.

Examine the section with a high **power**, studying **each tissue** seen in the transverse section.

14. The *epidermis;* compare the length of its cells with the same in transverse section. Draw a few cells.

15. The *sclerenchyma;* cells elongated with tapering ends. Note the pits and the middle lamella as in transverse section. The mouths of the pits may be seen when a wall extends across a cell. Draw.

16. The *cortical parenchyma;* as in transverse **section except that** the cells are elongated.

17. The *bundle-sheath;* the length and narrowness **of the** cells.

18. The *fibro-vascular* **bundle;** the two regions distinguished in transverse section, xylem **and** phloem. Commencing at the bundle-sheath notice

    a. The *phloem parenchyma;* **much** as in the former section.

    b. The *sieve cells;* their great elongation, tapering **ends** overlapping succeeding ones, and slightly **thickened** walls. Note the sieve plates on **the side** walls, looking **like** irregular thin spots with fine specks in them; or the sections of them on the **cut** edges of the vessels, as depressions of the surface of the **wall, paired,** one **on each** side when **two** sieve vessels **are** contiguous.'

    c. The *scalariform vessels.* Observe

' Under a ⅛ or ₁⁄₁₀ objective the pores in the sieve plates may be better seen. Consult Bessey, Botany, fig. 71, p. 81.

i. That the walls of these vessels present many narrow thin spaces, looking like slits placed transversely.

ii. That these thin spaces do not extend entirely across the face of a vessel.

iii. Find a place where the contiguous walls of two vessels have been cut through by the razor and observe the beaded appearance of the walls. Each "bead" corresponds to the thick part of the wall and the intervals to the thin places.

Isolate some scalariform vessels by boiling a rather thick longitudinal section for a few seconds in potassic chlorate solution. Mount in water and examine with a high power [8]

iv. The shape and markings. Draw.

d. Draw the fibro-vascular bundle with a portion of the bundle-sheath.

19. The *trichomes,* in the form of scales. Mount scales of various shapes under the same cover, and observe

a. The shapes and arrangement of the cells, especially at the apices of the scales.

b. Draw a scale.

B. THE ROOT. Cut a transverse section of one of the larger roots, examine with a high power, and note

1. At the edge of the section (if perfect) the *epidermis,* [9] of irregular thick-walled cells, not differing much from

2. The underlying brown *parenchyma,* which gradually merges into

---

[8] If the vessels do not separate in mounting, press gently on the coverglass with a little sidewise push.

[9] So called here because of its position ; not necessarily homologous with the epidermis of the stem.

3. Yellowish *sclerenchyma*, similar to that of the rhizome. Notice the starch, increasing in quantity **toward** the center.

4. **Draw** a portion of the above tissues.

5. Note the abruptness with which the sclerenchyma **joins**

6. The *fibro-vascular bundle.* Notice the **sheath** which encircles it.[10]

    a. Just within the bundle-sheath, a row of parenchyma cells with granular contents (protoplasm) er.circling the bundle, the **pericambium.**

    b. The *xylem region ;* consisting of

        i. *Scalariform vessels ;* four (sometimes three or five) of which, occupying the center of the bundle, are in pairs, one pair larger than the other ; the remainder, much smaller, **are in two** clusters, one between **the vessels of the** smaller pair and the **pericambium on each** side. If the scalariform vessels are not easily made out, a section may be treated with potash or stained with iodine. They then become very plain.

        ii. *Xylem parenchyma;* packed between and immediately around the larger vessels.

    c. The *phloem* **region ;** its **two** parts separated by **the** xylem, lie **outside of** the vessels of the larger pair and consist **of** parenchyma **with** granular contents and empty sieve vessels.

7. Considering the **whole** bundle, **notice** that all the tissues it contains are symmetrically disposed about a center. It is therefore known as a **radial bundle.**

---

[10] The bundle sheath frequently breaks in cutting.

8. Draw the bundle.

Take one of the largest roots whose root-cap is present and cut a series of longitudinal sections, mount, treat with potash, and selecting the section which passes through the center of growth, note

9. The concentric layers of the *root-cap*, each thickest in the middle, the outer sloughing off.

10. The tissues at the apex of the root. In the center, immediately under the root-cap, a large triangular cell, apex inward, the apical cell. Notice that the cells adjacent to the inner faces of the apical cell have evidently been derived from it by partitions parallel to its faces.

11. Draw the tip of the root, including the root-cap.

12. The trichomes in the form of *root-hairs*. Slice off from a root a thin piece carrying a number of hairs, and note

    a. Their *attachment* to epidermal cells.

    b. The *shape* of a hair near the proximal and distal ends.

    c. The *color* of the wall and absence of septa.

    d. The occasional *spiral thickenings* in the large hairs, usually forming a loose spiral of three or four turns only.

    e. The *contents*.

    f. Draw a hair showing these points.

C. THE LEAVES.

1. The *epidermis*. Lift the epidermis of the lower surface of a leaflet with the point of a needle, seize it with fine forceps and strip off as much as possible, mount, examine with high power, and notice

a. The very irregular shape of the cells and the way in which they dovetail into each other.

b. Here and there narrow slit-like *stomata*, each bounded by two crescentic cells, the *guard cells*.

c. Along certain lines (over the veins) the different shape of the cells.

d. The *chlorophyll bodies*, especially in the guard cells ; their granular nature.

e. Make a drawing showing these points.

f. Examine in the same way the epidermis of the upper surface of a leaflet ; note the absence of stomata.

Cut a vertical section of a leaflet at right angles to the veins. Observe

g. On each side of the section the irregular epidermis, containing chlorophyll bodies. On a drawing of the surface view of the epidermis draw imaginary lines in various directions and note the differing lengths of the lines across any cell. This will explain the different lengths of the epidermal cells cut by the razor.

h. Occasionally a stoma in the epidermis, bounded by the two guard cells, communicating with an intercellular space ; note the shape of the guard cells.

2. Occupying the space between the upper and lower epidermis, the loosely arranged irregular parenchyma of the leaf, *mesophyll*, also containing chlorophyll.

3. The large *intercellular spaces* of the parenchyma.

4. At intervals along the section the cut **ends** *of the veins.* Identify the tissues with those seen in the stem.

5. Beneath the vein, forming a part of the lower surface

of the leaf, will be seen three or four very thick-walled cells.

6. **Draw** the vertical section of the leaf.

Bend a leaflet over the finger and cut the thinnest possible slice from the under surface lengthwise of the veins. Mount with the cut surface upward, and note

7. The length of the cells over the veins and the manner of overlapping, **fibrous tissue.** Draw.

8. The trichomes in the form of *sporangia.* Scrape some sporangia from a sorus of a dried specimen, examine dry with a low power, and note

  a.   The *shape* and *color.*

  b.   The row of brownish walled cells extending part way around the sporangium, the *annulus.*

  c.   The *stalk* by which they were attached.

Examine with a high power sporangia from alcoholic specimens, mounted in water with cover, and note

  d.   The structure of the *wall* of the sporangium. This can be best studied in some of the immature sporangia which can usually be found in the same sorus with the mature. Supplement this study by examining the wall of a bursted sporangium. Observe that the wall consists of a single layer of much flattened cells. Note the nuclei.

  e.   The *annulus.* Study the cells which compose it. Notice that it forms a distinct ridge and is continued beyond the point where the cells are thickest by a series of short broad cells with thinner walls.

  f.   The *stalk ;* the number of cell rows which compose it and the absence of any trace of a fibro-vascular bundle.

g. Draw a sporangium.

h. The place of *attachment.* Scrape away most of the sporangia from an indusium, mount it, and notice the place of attachment of the remaining sporangia and of the bases of the stalks of the others. Observe its relation to the vein.

i. The mode of *dehiscence.* Tear off a bit of indusium bearing a few sporangia, from a dried specimen previously soaked in water, place on a slip without a cover glass and allow it to dry while examining it with a low power, illuminating it from above. Watch the process of bursting carefully.

j. The *spores.* In unbursted sporangia from alcoholic specimens notice how closely they are packed. Examine some which have escaped, and note

   i. Their *shape* and *contents.*

   ii. Their *double walls.* Burst some spores by pressing on the cover. In favorable specimens the outer layer of the wall, *exospore,* will be ruptured and the delicate inner layer, *endospore,* with its inclosed protoplasm will be seen protruding.

D. THE PROTHALLIUM. Carefully brush away all the dirt from the under side of a prothallium of medium size and mount it with the under side uppermost. Examine with a low power, and notice

1. The *shape* and the character of the *margin.*
2. The shapes of the *cells.* Draw a few cells of the prothallium showing the various shapes.
3. The abundant *chlorophyll bodies.*

4. The absence of fibro-vascular bundles.

5. The *trichomes* in the form of hairs of various kinds.

   a.   Shorter or longer pointed hairs on the surface and margin.[11]

   b.   Short blunt hairs in like positions.

   c.   *Rhizoids.* Note

      i.   Their various sizes and lengths.

     ii.   The irregular shape. Draw.

6. Roundish bodies of considerable size near and among the rhizoids, the *sexual organs.*

   Examine with a high power, and notice the two sorts :

   a.   Some bodies spherical and filled with smaller cells, the *antheridia.* Observe

      i.   The single layer of cells forming a wall [12] which incloses

     ii.   A cluster of spherical cells, the sperm cells. [13]

    iii.   If fresh material is being used, some mature antheridia will probably have been ruptured in mounting and the sperm cells with their *antherozoids* have escaped. Note the *movements* of the antherozoids after they have escaped from the sperm cells. Kill them by treating with iodine, watching them as they come under its influence. Take note of the *body* of the antherozoid, a spirally coiled filament to which is usually attached an almost *empty*

---

[11] Sometimes wanting.

[12] Best seen in immature antheridia. All stages may usually be found on the same prothallium.

[13] If the structure of the antheridia can not be made out easily here, postpone the study till D. S. *a.* is reached.

vesicle, and of the numerous *cilia* [14] **at the free** end of the body. **Draw.**

iv. If no fresh prothallia are **procurable the** antherozoid may **be seen** within **the sperm** cell in alcoholic specimens, but **its parts are not** distinguishable.

v. **Draw both** young **and** mature **antheridia, showing** structure and contents.

b. Some bodies, of similar shape to antheridia but apparently composed of four cells either quadrant-shaped and meeting in the middle or somewhat oval leaving a squarish **space** between them, the *archegonia.*

i. In **favorable fresh** specimens one or more **moving** antherozoids may **be seen in** the space, *canal,* between the **four cells.**

Cut several vertical sections of the prothallium, passing through the region of the notch and the cluster of rhizoids. Treat with potash, examine with a high power, and notice

7. The number of cells in thickness of various parts of the prothallium ; especially its rapid thickening in **the region** of the rhizoids.

8. The *sexual organs.* [15]

a. The **globular** *antheridia,* wholly superficial. **Notice the thickness** of the wall in mature and immature ones.

b. The *archegonia* **may** be recognized **by the more or** less recurved projecting neck composed of **several rows of cells.** Note

---

[14] Difficult to see. Use ⅛th **or** higher objective if possible.

[15] If their structure has not been comprehended before, it may be easily **made** out now by examining numerous sections of the prothallium.

i. The number of rows of cells composing the *neck*.

ii. The *canal* between the cells of the neck, and extending from its apex to the imbedded portion of the archegonium. This canal is difficult to distinguish unless it contains a granular substance.

iii. The cluster of cells at the base of the neck imbedded in the prothallium, the *body* of the archegonium.

iv. At the inner end (base) of the canal, in the midst of the cells of the body, a single large central cell, filled with a rounded mass of protoplasm, the *oosphere*.

v. Draw the archegonium.

ANNOTATIONS.

Regarding only the position of organs, perhaps the most striking difference between Adiantum and Atrichum is to be found in the fact that the former has its leaves only above the ground, while the real stem is buried below it. In contrast with those low plants whose rhizoids have served them well enough for holdfasts, the fern has developed strong fibrous roots which ramify widely and perform this office, assisted by the buried stem. These roots are made necessary not only by its greater stature and the consequently greater strains, but by the necessity of wider foraging for the supply of food. The roots must push their way among the particles of soil, and, to protect the tender tissues of the growing point, the tip of the root is covered by a cap of cells, which arise from segments cut off from the outer face of the tetrahedral apical cell.

As the cap is gradually disorganized and worn away by contact with the soil it is replaced by new growth from behind. The root cap is to be considered as a modified and augmented portion of the epidermis.[18]

Provision for continued growth of the stem in length is found in the bud at its apex. The dying base, however, follows with equal pace the advancing apex, severing the lateral branches as it reaches them, which thus become independent plants.

One of the most marked advances upon the structure of the moss is to be found in the development of an extensive and complicated fibro-vascular system. The simple leaf traces of Atrichum are here replaced by better developed groups of fibers and vessels to which the term fibro-vascular bundle is applied. These bundles are distributed to every part of the plant; condensed in those parts requiring strength, such as the roots, stem and leafstalk; diffusely branched in the leaflets for the support of the chlorophyll-bearing tissue. Branches of the fibro-vascular bundles having once been formed, do not reunite with their fellows, either as a whole or by anastomosing branchlets. The only organs of Adiantum not reached by the fibro-vascular bundles are the numerous and unusually varied trichomes. These are developed as scales thickly clothing the stem and base of the leaf stalk, as hairs matted together about the roots, and as sporangia crowded under the edges of the leaflets.

In the growing parts of all organs of the fern, the cells are parenchymatous, but certain groups early dif-

[18] Bessey, Botany, p. 163.

ferentiate into the tissues which compose and surround the fibro-vascular bundles. These tissues are quite distinct from each other as well as from the original parenchyma.

The sheath which incloses the bundles does not belong to the bundle itself, either in the fern or other plants, but to the surrounding parenchyma.

The apparently perforated plates on the walls of the sieve cells can not be seen clearly because of the layer of protoplasmoid substance which adheres to the walls. The perforations themselves are not easily demonstrated though DeBary [17] thinks he has seen fine filaments connecting granules on opposite sides of a plate. The continuity of protoplasm between other than sieve-cells has been demonstrated in many plants.

The arrangement of the tissues of the bundles in stems and roots is of different types. In the former, the phloem of the bundle encircles the xylem whence it is known as a concentric bundle. In the latter, the xylem forms a plate dividing the phloem into two portions which stand one on each side of it. Assuming a center, the xylem and phloem masses are symmetrically disposed about it, whence the bundle is known as radial.[18]

The root-bundle contains a tissue, the pericambium, whose cells are still capable of division; no such tissue is found in the stem-bundles. New roots have their origin not in the pericambium as in phanerogams, but from cells of the bundle-sheath.[19]

---

[17] Comparative Anatomy, p. 181.

[18] Cf. Strasburger, Das botanische Practicum, p. 209; DeBary, Comparative Anatomy, p. 362.

[19] Cf. Strasburger, Das botanische Practicum, p. 276; Prantl and Vines, Text-book of Botany, p. 51.

The original parenchyma outside the bundles of the
stem early thickens its walls. These thick walls consist of several layers, the most prominent of which, the
median, is called the middle lamella. This layer,
according to Strasburger [20] and others, is the primary
cell wall, upon which thickening layers are deposited.
By other histologists it is held that the layers are
formed, as the thickening progresses, by the differentiation of the wall. Growth in thickness, according to
the first view, is due to apposition; according to the
second, to intussusception.

The thickening layers of the wall are perforated by
numerous pits, through which probably pass threads of
protoplasm, not occupying the breadth of the pit, but
passing through much more minute openings in the
closing membrane of the middle lamella.[21]

In addition to serving to increase the strength of
the stem, the cortical part is a convenient storehouse
for reserves of food, as indicated by the quantity of
starch in its cells.

The several cell layers of the leaf necessitate some
arrangement for allowing the entrance of gaseous food
and exit of the by-products of the cells' activity;
hence the loose arrangement of the cells of the leaf,
forming large intercellular spaces, which communicate
with the exterior by numerous stomata. The stomata
have here the form usual among the higher pteridophytes and flowering plants, an elliptical slit, bounded
by two crescentic cells, which by their change of posi-

---

[20] Bau und Wachsthum der Zellhäute, p. 175.

[21] Cf. Schaarschmidt, **Protoplasm**, Nature, xxxi, p. 290; Gardiner,
ibid, p. 390.

tion may either open more widely or almost close the orifice.

The prothallium, which is developed from a spore produced by the leaf, bears little resemblance to the mature spore-bearing fern plant. In its flattened shape, cellular structure and rhizoids it does, however, have a striking resemblance to the thalloid stem of Marchantia.

There are thus two distinct stages in the life history of the fern: one is known as the vegetative, asexual or pteridoid stage, in which the plant consists of stem, roots and leaves, and produces spores, and, strangely enough, answers to the sporogonium of the moss; the other, known as the reproductive, sexual or thalloid stage,[22] in which the plant consists of a prothallium, on which the reproductive organs are borne, and corresponds to the leafy plant in the moss.

These reproductive organs are quite like those of Marchantia and Atrichum. The antheridia consist originally of one cell, which is later cut up into a central cell and several parietal ones. The contents of the central cell are divided into a number of small spherical cells in which are formed the antherozoids. When these are mature the parietal cells absorb water and burst the apical one, thus permitting the antherozoids to escape. The body of the antherozoid according to Strasburger[23] is to be regarded as the protoplasm of the nucleus of the sperm cell, and the cilia as

---

[22] Pteridoid and thalloid are terms introduced by Underwood, Our native ferns and their allies, p. 35.

[23] Das botanische Practicum, p. 455; Sachs, Text-book, 2nd Eng. ed., p. 423.

the peripheral protoplasm of the cell. The vesicle attached to the hinder coils of the body is formed from the central or intermediate contents of the sperm cell, and usually contains some starch grains.

The archegonium is likewise originally a single cell of the prothallium, which by subsequent division forms a central cell containing the oosphere, the two canal cells whose destruction results in the formation of the canal, the four rows of neck cells and the layer of cells immediately surrounding the central cell. "

The conversion of the two canal cells into mucilage, and the partial expulsion of this from the canal, entangles and allows the entrance of the antherozoids, which by their active movements work their way to the base of the canal and penetrate the wall of the central cell in which lies the oosphere. One antherozoid bores its anterior end into the germinal spot of the oosphere and disappears within it, probably reaching the nucleus. The others lie for some time upon the oosphere and are gradually absorbed, only one antherozoid actually penetrating it."

The result of the fertilization of the oosphere is the formation of a new plant, which remains attached to the prothallium on its under side for some time. As the young fern gradually spreads sufficiently, and is able by means of its leaf and root surface to gather nourishment for itself, the prothallium, no longer useful, perishes.

---

[74] Cf. Sachs, l. c.

[75] Strasburger, op. cit., p. 458.

SCOTCH PINE.

Pinus sylvestris L.

PRELIMINARY.

THE Scotch pine is a species commonly planted for ornament. It may be readily recognized by the following characters. At a short distance the tree has a grayish-green color. The leaves are in pairs, five to ten centimeters (two to four inches) long, somewhat twisted, covered with a whitish powder which can be rubbed off with the fingers and to which the peculiar color of the tree is due. The cones are small, about five centimeters (two inches) in length, the free ends of the scales being produced into conspicuous protuberances, which near the base of the cone are recurved.

The Austrian pine, a two-leaved species also commonly planted for ornament, differs from the preceding in having longer leaves—from ten to fifteen centimeters (four to six inches) in length—with a dark green color without any of the powder. The cones are much larger and without the recurved protuberances. If the Scotch pine can not be procured the Austrian will do quite well, being closely similar to it in structure.

The flowers, both male and female, should be collected in spring as soon as the male flowers begin to scatter their pollen. The male flowers when mature form conspicuous yellow clusters at the base of the young shoots. The female flowers are quite inconspicuous, in small oval clusters of a pinkish color, projecting slightly beyond the ends of the young shoots. The tree bearing abundant male flowers usually bears few female ones, and vice versa. These flowers when collected should be preserved in alcohol. A few weeks later the two-year-old cones, which will be found just below the new shoots, should be collected and preserved in alcohol. If the plant is to be studied in spring or summer, some of the large terminal buds should be collected in the late autumn, winter or early spring preceding, and preserved in alcohol. Leaves and stems should be gathered about the first of July, and preserved in alcohol. Mature cones should be gathered in winter or early spring and allowed to dry, care being taken to prevent losing the seeds, which will shake out on drying.

Fresh leaves and stems may be used for the study of the gross anatomy, but if used for the minute anatomy it is well before cutting sections to place them in alcohol for a few days to get rid of the resin which exudes and gums the fingers and knife unpleasantly. Before cutting sections of stems or leaves which have been preserved in alcohol and before dissecting the male and female flower clusters, it is well to place them for a day in a mixture of equal parts of alcohol and glycerine, which renders them somewhat easier to manipulate. They may, however, be used direct from the alcohol.

The requisites for the complete study are stems,

leaves, terminal buds, male and female flowers, **year-old** and two-year-old cones, preserved in alcohol; mature cones and seeds, dry; alcohol; potash; glycerine; sulphuric acid; and if convenient, **magenta**; methyl blue; and chlor-iodide of **zinc**.

LABORATORY WORK.

GROSS ANATOMY.

A. GENERAL CHARACTERS. **Note**

1. The central **axis** or *stem ;* its few *main branches* and numerous very short **dwarf branches** [1] bearing

2. Pairs of very slender elongated green **needle leaves.**

3. **The scales upon the stem, those** covering the buds at the apex of the stem and those overlapping the bases of young leaves. All may be called *scale leaves.*

4. Near the base of the young shoots in some specimens, a number of oblong (nearly globular) **clusters** of light yellow bodies, **stamens**, the *male flowers ;* in other specimens, **one** or **two** small oval clusters of *female flowers*, projecting beyond the end of the stem.

B. **THE STEM.** Examine

1. The *surface* of a year-old **shoot.** Note the scales covering it, especially near the base of the shoot. Compare with the surface of older **stems** ; note the gradual obliteration **of** the scales.

2. The arrangement of the *main brnches.*[2] Note the

---

[1] The terms "dwarf branches" **or** "dwarf **shoots**" will **be** used to distinguish these from the main branches **or shoots.** (The term shoot includes the branch with **its leaves.**)

[2] **Best** seen in specimens from young vigorous **trees.** If possible the student should study the tree itself.

number of branches and the relative vigor of terminal
and lateral shoots. Compare also, as to size, the *buds*
found in clusters at the apices of the branches.

3. The arrangement of the *dwarf branches*. Select the
straightest and most vigorous year-old branches for
this study. Notice

a. The *position* of the branches relative to the scales.

b. Their absence from certain portions of the stem.

c. **Pull** out the pairs of leaves from fifteen or twenty
consecutive branches. Stick a pin at the base of
any branch, and then find a branch that stands
directly above this one. Count the number of
branches between these, including the first. This
number will be equal to the number of vertical
ranks in which the branches stand.

d. Make a *diagram* in the following manner, to show
the relative position of the branches : draw lightly
a number of concentric circles about three millimeters apart (the number should be twice as
many as the number of vertical ranks, plus one).
Divide the outer circle by as many equidistant
points as there are vertical ranks of branches.
From these points draw radii, lightly. Take a
piece of straight stem about ten centimeters long
which has been stripped of its leaves. Mark the
position of three or four consecutive branches by
pins, so placed that if pressed in they would pass
through the center of the stem. Fasten the lowest
pin securely. Make a mark on the outer circle at
any radius to indicate the position of the branch
marked by the lowest pin. Erect the stem at
the center of the circles, making the lowest pin
coincide with this radius, and mark the next

higher branch on the second circle at the radius with which its pin now most nearly coincides. Mark **the** third and fourth in the same **way.** Leaving **the lowest** pin in place, move the pin next lowest to the **next** higher **unmarked** branch, **and mark** its **position.** Repeat **this until all the** circles are filled, numbering each branch from the lowest up. Studying this diagram determine

i.   The *arithmetical difference* between the numbers of the branches which lie on the same radius.

ii. The *number of turns* made by a spiral line joining successive branches, 1, 2, 3, 4, etc., until it reaches a branch over the first.

iii. Find a fraction which will express the part of a circle intervening between any two suc- cessive branches.

iv. Note that the numerator of this fraction ex- presses the number of turns made by the spiral line, and the denominator the number of ranks in which the leaves stand.

4. The *buds.*   Notice

a.   Their *position* and relative *size.*

b.   Their *shape.*

c.   Their *structure.*   Study

i.   The *scales.* Carefully strip them from the bud with needles. Note particularly the character of the *edges* and the differences between the *apical* and *basal portions.* After removing the brown apical portion, the green basal parts will be seen closely investing

ii. The *axis.* Bisect longitudinally the portion of the bud remaining. Observe in the center the whitish stem or axis, tapering gradually and then rapidly to **a** point, and bearing the thick-set bases of the bud scales, **in** the **axils** of which may be seen

   iii.   *Secondary buds.*[3] Take out one of these buds carefully and dissect it. Note the *scales* which cover it. By cautiously removing these **the** *rudimentary needle-leaves*, looking like **two** minute knobs, may be found, apparently at the end of a very short stem to which the scales were attached.

   iv.   Make drawings showing the external appearance and structure of the buds, both main and secondary.

  d.   Compare the buds with the branches. **Observe** **that** a **bud** is simply an undeveloped **branch.**

5. The *structure.* Cut an old stem square across to study the cut surface. **Mount also a** transverse **section of** the same. Notice

  a.   A central yellowish **or** brownish spot **of** irregular outline, the **pith.**

  b.   Surrounding the pith a zone of firm tissue, the **wood.** Observe

   i.   **The** concentric masses of tissue, **growth rings,** **the** number depending upon the age of the **shoot at the** point cut. In thin parts of the **section,** notice the difference between the **central and** peripheral portions of **any growth** ring.

   ii.   The many **fine** radiating **lines, the medullary rays. Note the** extent **of the** larger ones.

   iii.   Many small scattered openings, **the resin ducts.**

---

[3] These can only be found of sufficient size to dissect in buds collected late in autumn or in early spring just before they begin to expand.

    iv. In some sections one or more distinct whitish *bundles* passing out from the center of the stem. Notice that a continuation of the central pith occupies the center of each. Observe the relation of these bundles to the scars on the bark indicating the position of former dwarf shoots. If the stem be four or more years old, note that the bundles stop quite abruptly at the close of the second year's growth.

  c. All the part outside the wood, the **bark.** Distinguish its three layers :

    i. The inner *fibrous layer*, whitish. Notice its appearance and thickness relative to the whole bark.

    ii. The middle, *green layer*. Notice the large resin ducts. (In fresh specimens note the color, consistence and odor of the liquid they exude.) Compare the thickness of this layer with that of the first.

    iii. The outer *brownish layer*, except in quite old stems made by the adherence of the bases of the scale leaves. Note its relative thickness.

    iv. Strip off a portion of the bark. The three layers may be easily separated with the fingers. Study the characteristics of each.

  d. Bisect the stem longitudinally. On the cut surface and in thin sections make out the pith, wood and bark ; the growth rings, medullary rays, and bundles extending toward bases of former leaf-branches, in the wood ; the three layers of the bark.

  e. Make drawings of the transverse and longitudinal sections to show completely the structure of the stem.

6. **The *dwarf shoots*.** Carefully break one from the stem, and note

   a. The *scales* (scale **leaves)** enwrapping it and the bases of the needle leaves. If possible **compare** these scales on young and old dwarf shoots.

   b. The *length*.

   c. The **very** small rudimentary *terminal bud* between **the** leaves. This is best seen on the dwarf shoots from young vigorous **trees. It is minute or** absent on others.

C. THE LEAVES.

1. **The *scale leaves*. These have** already been studied as they occur on the dwarf shoots (B. 6. *a*.) and in **the bud (B. 4. *c*. i.). Compare the scales of** the stem with those of a young bud and notice the **loss of** the deciduous **apex.**

2. The *needle leaves*.   Note

   a. The *number* on each dwarf branch.

   b. The *shape* and *apex ;* also the shape of the transverse *section*.   Draw a leaf.

   c. The *color*. Compare old and young leaves if **possible.**

   d. The *texture ;* **firmest** near the apex, softer near the base, due **to** *basal growth*. These points are especially noticeable in young leaves.

   e. The *edges*. Draw the finger from the apex toward **the base.** Examine with a lens.

   f. The *surface*. Observe

      i. That **it is** faintly whitened (*glaucous*) **by a** powder which can be removed by drawing the leaf through the fingers ; best seen on the flat side.

  ii. The longitudinal rows of whitish dots on both surfaces. Cut a thin slice from the convex surface, mount, and examine by transmitted light. If sufficiently thin, the dots will now be seen to be minute openings, the *stomata* or breathing pores.[4]

 g. The *structure.* Cut a transverse section and examine by transmitted light. Notice

  i. Occupying the center an oval patch of whitish tissue, the *fibro-vascular region.*

  ii. Outside the central whitish area, compact green tissue, *mesophyll.* In this zone notice a dozen or more openings, the *resin ducts.*

  iii. Enveloping the whole, the narrow colorless *cortical area.*

  iv. Draw the section.

Cut a longitudinal section parallel to the flat side. Make out the same regions as in the transverse section.

D. THE FLOWERS.

 1. Male or *staminate.* Carefully break off one of the clusters.

  a. Note the short *stalk* by which it was attached to the stem.

  b. Note that the cluster is made up of numerous short-stalked bodies, the *stamens*, attached to an axis. Each stamen consists of a flat scale bearing on the inferior surface two enlargements, the pollen sacs.

  c. Burst a pollen sac. Note the innumerable minute grains of pollen which escape.

[4] More accurately, the external chambers of the stomata, for the real stomata are deep seated.

d. Find a stamen which has burst spontaneously. **Note** how it is ruptured (by slits).

e. Note the arrangement of the *flowers*.[5] They are almost sessile and crowded on an elongated **axis** forming a **spike**. Notice the *scale* subtending each flower, and the number and position of **the scales** attached to the short stalk of the flower.

f. Note the *position* of the flowers ; each replaces a branch on the young shoot.

g. Draw a stamen showing its structure.

2. Female or *pistillate*. **Taking a** single cluster, a *spike*, notice

a. The stalk, **peduncle**, by which it is attached ; **its** *direction ;* the *scales* on the peduncle.

b. That it **is composed of two kinds** of scales : (1) thin, the **bracts** ; (2) thick, the **carpellary scales.**

Dissect out a single *bract ;* **note**

i. The *texture* and *shape*.

ii. Draw the bract.

Dissect out a single *carpellary scale ;* note

iii. The *shape* and *texture*.

iv. The prominent **keel** on **the** upper surface in the median line.

v. Two enlargements on the superior **surface,** near the proximal end, **the ovules.** Notice the *position* of **the** ovules and the large ori- fice **at** their free ends, the **micropyle, the** *integument* of **the** ovule being prolonged into a short **tube, whose** right **and** left sides are still further produced into **two short** fila- ments.

---

[5] It is assumed that each cluster of stamens constitutes a single flower

    vi.   Draw a scale showing all these points.

  c.   Difference in the *size* of bracts and scales in **different** parts of the same cluster.

  d.   Position of the cluster ; replacing **one of the main branches.**

Examine a year-old cone. Bisect it vertically, and note

  e.   The central tapering axis.

  f.   The cut edges of the scales and bracts. Observe the relative thickness of the scales at their proximal and distal ends.

  g.   The ovules appearing in section at the base of the scales.

  h.   Whether the scales are free from each other or adherent.

  i.   Draw the cut surface.

Dissect out a scale with its ovules. Notice the many scales with abortive ovules. Bisect a well developed *ovule* carefully, through the micropyle. Note

  j.   The diminished size of the *micropyle.*

  k.   The single *integument.*

  l.   That portion inclosed by the integument, the **nucellus.**

  m.   Nearest the base of the nucellus (the end nearest the micropyle being considered the apex) a large cavity, the **embryo-sac,** partially or wholly filled with a soft substance, the **endosperm.**

  n.   Draw the cut surface of the ovule.

E. THE FRUIT (CONES). Examining a mature cone, notice

  1.   The large *carpellary scales,* making the bulk of the cone. Observe their *color,* above and below, *consistence, shape* and *markings* at the free ends.

  2.   In an open cone, or by cutting away the basal third of

a closed cone, the *smaller bracts* subtending the carpellary scales.

3. Closely applied to the superior surface of the carpellary scales, a pair of thin wing-like scales, each bearing at its proximal end a perfect **seed** or an abortive *ovule.*

4. The *seed.* Note

    a. The *shape, surface* and *markings.*

    b. At the pointed end notice the minute opening, the *micropyle.*

    c. The *structure.* **Bisect** the seed longitudinally parallel with the flatter faces, and in the halves make out

        i. The firm *coat.*

        ii. The inclosed portion consisting **of two** parts : (1) the young plantlet, **embryo,** lying in the **axis**; (2) the food **for the** plantlet, *endosperm.*[8]

        iii. Note the *position* **of** the embryo with respect to the micropyle.

Take another seed and with needles dissect off

        iv. The *coat.* Notice that it has differentiated into two layers. Compare the two as to color, thickness and strength.

Dissect the endosperm carefully from th.e *embryo.* In the latter make out

        v. The short stem, **caulicle.**

        vi. The six divisions arising **from** about the apex of the caulicle, the first leaves, **cotyledons.**

        vii. A minute elevation in the midst of the cotyledons, at the **apex** of the caulicle, the rudimentary terminal bud, **plumule.** (Not easily seen.)

---

[8] The endosperm has therefore entirely displaced the nucellus originally surrounding it. (See D. 2. *l.* and *m.*)

MINUTE ANATOMY.

A. THE STEM. Cut a transverse section of a year-old stem, examine with a low power and note

1. The *pith*, occupying the center of the section. Observe

   a.   The *outline* of the pith.

   b.   In some sections a portion extending outward to enter a dwarf branch. The salient *angles* of the pith are all due to such outward extensions at different heights.

   c.   The loose arrangement of its cells.

2. The *wood* (*xylem*). Observe

   a.   The arrangement of the cells.

   b.   The openings of the *resin ducts*.

   c.   The division into two zones, *growth rings*.

3. The cambium ; a narrow, cloudy looking zone, bounding the xylem. (If the section be from a stem gathered in winter or early spring, the cambium zone will be indistinguishable.)

4. The *phloem* ; of compactly-arranged cells, with a whitish appearance.

5. The *cortical parenchyma* ; outside the phloem, consisting of large, loosely-arranged cells, which in sections of a fresh stem contain much chlorophyll. In this region note the large oval openings of resin ducts.

6. Dark lines from the pith to the cortical parenchyma, the *medullary rays*.

7. The edge of the section. The cortical parenchyma is bounded by a row or two of small close-set cells. All the tissue beyond this belongs to the bases of the scale leaves, which cover the stem.

Examine with a high power and study

8. The *pith parenchyma*.   **Note**
   a.   The *shape* and *arrangement* of the cells ; the modified shape of those passing out to a dwarf branch.
   b.   The *contents*.   Test with iodine.
9. The *xylem*.   Notice that the salient angles of the pith divide it more or less completely into *wedge-shaped bundles*.   Studying one of these wedges, note
   a.   At the apex one or two *resin ducts*.   Study their structure, noticing
      i.   The *shape* of the opening.
      ii.   The circle of rather delicate cells lining the duct, the *secreting layer*.   Note the granular **nucleus** in each, nearly filling the cell.
      iii.   The quite irregular circle of flattened **cells**, **with** longer diameters parallel with **the circum**ference, bounding the duct, the *sheath*.
   b.   Between the resin duct and **the** pith, forming **the** point of the wedge, a group of several *spiral* and *reticulated vessels*.   These are rather difficult to distinguish from the wood cells.   They may be recognized by their slightly thicker walls, the smaller diameter and rounder shape of their cavities.   On staining the section slightly **with** magenta, they take a somewhat deeper **color than the wood cells.**   After the section has lain for some time in glycerine they may be recognized by their greater opacity.
   c.   Forming the bulk of the xylem, the *wood cells* or *fibers*.   On account of the similarity of the markings (to be studied later) on their **walls** to those on tracheæ or vessels, they are called **tracheïdes**. Note

i.   *Shape* and *arrangement.*

ii. Their *emptiness.*

iii. Their thick *walls,* showing in thin parts of the section, a *middle lamella.*

iv. In the thinnest part of the section, search for places where the radial walls [1] of contiguous cells bow away from each other like two watch glasses placed with concavities together. They are most readily found in the youngest part of the xylem. In the most favorable sections these bowed walls may be seen to be interrupted at their points of greatest diverg- ence thus ⸺⊂⊃⸺. These are sections of the **bordered pits** (further described at A. 18. *b.* iii.).

v.   Compare the tracheides of the outer growth ring with adjacent ones of the inner one.

vi. Wide one-sided bordered pits where the tracheides adjoin the cells of the medullary rays.

10. The *cambium.* Note

a.   The radial rows of rectangular,[8] very thin-walled cells, passing abruptly on the one hand into the xylem, but shading almost imperceptibly on the other into

11. The *phloem.* Note the two elements which compose it :

a.   Angular thick-walled cells with a whitish luster and constituting the greater part of the phloem, the *sieve cells.* In favorable sections the radial

---

[1] *I. e.,* those lying along a radius of the stem.

[8] Very apt to be distorted in cutting.

walls of some of these cells will be found **perforated** by clusters of very fine pits, **looking like** fine parallel lines passing across the **wall. These are** sections of the *sieve plates ;* they occupy **the** same relative position as the sections of the bordered pits **of** the tracheides. **Note** the shape **of** the **sieve** cells next the cambium and next **the** cortical parenchyma.

b. **Near the** periphery of the **sieve** tissue an interrupted row of cells with brown or yellow contents in which are strongly refringent crystals. Near the cambium a similar row of cells, larger and rounder than the sieve cells and with colorless or slightly yellowish homogeneous contents, in which **a** small crystal **or** two may sometimes be seen. **These two broken rows of cells** are the *phloemparenchyma.*[9]

12. **The** *cortical parenchyma.* **Note**

   a. The *shape, size* and *arrangement* of **the cells. Compare** with **pith** parenchyma.

   b. The *contents.*

   c. The very large *resin ducts.* Compare their structure with those of the xylem (A. 9. *a.*). Note the cells of the sheath, larger, thicker-walled and not flattened as are those surrounding the ducts in the **xylem; the** secreting cells, similar to but more **numerous** than **those** of the **xylem ducts.**

13. The *medullary* **rays. In a** thin **part of** the section note

   a. Their *extent,* **from** pith to cortical parenchyma.

---

[9] Can be brought out by staining with chlor-iodide of zinc and better still by methyl blue.

b.   The shapes of the **cells in** the xylem and the gradual **transition** into the **cortical parenchyma.**

c.   The *contents* of the cells.

14. The *bases of the scale leaves.* (As they are closely attached to the stem, and the lower portions not distinguishable from it, their transverse section is most conveniently studied at this time.) Note the two layers :

   a.   The inner ; cells very thin-walled and irregular, apt to be distorted in cutting.

   b.   The outer ; composed of one or two rows of large cells, *sclerenchyma* (note shape), and a single outermost row of smaller cells, the epidermis. Note

      i.   The thickening of the outermost wall of the *epidermis.*

      ii. The continuous layer covering this wall, the *cuticle.*

15. Draw a part of the section, filling in sufficient to show the structure completely.

   Cut a longitudinal radial section of a year-old stem. Examine with a low power, and make out

16. The same areas as seen in transverse section, in this section appearing as strips :

   a.   The *pith ;* its regular margins.

   b.   The **xylem.** Note

      i.   Patches of transversely placed cells, the *medullary rays.*

      ii. The *resin ducts ;* showing as one or two lighter streaks in the xylem.

      iii. The two *growth* **rings.**

   c.   The *cambium ;* a very narrow whitish strip.

   d.   The *phloem ;* compact and fibrous-looking.

   e.   The *cortical parenchyma.*

f. The *scale leaves.*

Examine with a high power. **Study**

17. The *pith cells.* **Note**

    a. The shape and arrangement.

18. The *xylem.* Note

    a. Near the **pith** parenchyma a cluster of *spiral* and *reticulated vessels.* Notice the irregularity and closeness of the spiral thickening.

    b. The *tracheides*, making the bulk of the **xylem.** Note

        i. Their *shape.* Observe their **ends.**

        ii. Their thickened *walls.*

        iii. Their markings, *bordered pits.* In the youngest part of the xylem study the *structure* of one of these pits. Observe the **two concentric circles they present.** Note **which is more distinct. Compare with transverse** section **and** discover the **cause of this** appearance. **The** outer circle **is at** the point where, **in** section (see diagram at **A.** 9. *c.* iv.) the arms of the **Y** diverge from its stem ; the inner is the edge of the opening in the bowed walls. By examining this section thoroughly, chance sections of the pits may be found which will **further** elucidate their structure.

        iv. The *size* of the pits compared with the breadth of the fibers, **and** their arrangement on the fibers.

        v. The large thin spots on the walls **of the cells** of the medullary rays, where **they join the** adjacent tissues.

    c. Between the tracheides and the spiral vessels a few intermediate cells with *plain pits* nearly or

quite as large as the bordered ones of the tracheides. By focusing carefully the walls of these
cells may sometimes be seen in section.

19. The *cambium.* Note the *shape* and *contents* of the cells.
There is sometimes difficulty in discovering the end
walls of the cambium cells. It can be obviated somewhat by examining a section which has lain in glycerine
for a few hours. Notice particularly the delicacy of
the walls.

20. The *sieve cells.* Study
    a. Their *shape* and *arrangement.*
    b. The *markings* on their walls ; round or oval areas
    of fine perforations, looking like minute specks.
    Note their arrangement ; compare with that of the
    bordered pits on the tracheides.

21. The *phloem parenchyma ;* note length and contents of
the cells.[10]

22. The *cortical parenchyma.*
    a. Study the shapes and contents of the cells.
    b. Notice here and there cells which seem to have
    been divided by a partition, the pair still retaining
    an oval shape.
    c. The large intercellular spaces.

23. The *medullary rays.* Study their cells in the cambium
and sieve cell regions.

24. The *resin ducts.* (Their longitudinal structure may be
studied either in the longitudinal or transverse section
of the stem, the latter usually showing a longitudinal

---

[10] Difficult to distinguish without staining with methyl blue or chlor-iodide of zinc.

section of one or more of the horizontal branches connecting neighboring ducts. The structure is most
easily made out in those of the xylem, those of the
phloem being too large to allow a complete section to
be easily obtained.) Note

a. The empty cells forming the sheath ; their shape.
b. The secreting parenchyma cells lining the duct ;
    shape and contents.

25. The *bases of the scale leaves.* Note

   a. The delicate thin-walled cells forming their inner
       portion.
   b. The rather thick-walled cells, sclerenchyma, form-
       ing the outer part.
   c. The very thick-walled outer row, the epidermis,
       with thickly pitted walls.
   d. The very thick cuticle.
   e. The contents ; note color.

26. Draw a portion of the section, showing all the above
    points.

Cut a longitudinal tangential section, passing through
the wood. Examine with a high power, and note

27. The cut *ends of the medullary rays,* wedged between
    the fibers of the xylem. Notice

   a. The number of rows of cells in the thickness and
       height of each ray.
   b. The thin parts of the walls corresponding to the
       pits (see A. 18. *b.* v.).
   c. Make a drawing of one of the rays, showing also
       a few adjacent tracheides.

28. The numerous sections, in different directions, through
    bordered pits. Study these sections further, if necessary
    to an understanding of the structure of the pits.

29. The very tapering *ends of the tracheides.*

Cut transverse and longitudinal sections of a young stem collected at flowering time. Examine with a high power, and compare with similar sections of the older stem. Notice the walls and contents of the cells of the several tissues and particularly

30. The distinctness of the spiral and reticulated vessels.

31. The deep indentations of the margin of the stem in transverse section, marking the breadth of the scale leaves.

32. The simple epidermal and hypodermal tissues constituting the bases of the scale leaves.

Strip off the brown apical portions of the bud-scales from a winter bud and bisect it longitudinally a little to one side of the center. Cut a series of longitudinal sections as uniformly thin as possible, until the center of the stem has been passed. Mount every section, treat with potash and examine with a low power. Search for the section which includes the center of the axis. It may be recognized by the conical shape of the apex. Note

33. The central axis or stem.. Observe the arrangement of the cells.

34. The buds on the side of this axis. Notice

    a.   The large scale (base of bud scale) subtending each. .

    b.   The central rounded mass of cells, an undeveloped dwarf branch, covered in by scales. Search for a bud whose central part shows three rounded protuberances. These are the two leaves with the terminal bud of the dwarf branch between them. Draw.

35. The **conical** apex, *growing point*, of the axis. **Notice** the scales which cover **it**.

Examine with a high power. Study both the **growing** leaves and the apex of the stem. Note

36. The shape of the cells, which in these regions are capable of division and **are** collectively known as the **primary meristem.**

37. A short distance behind the growing apex, the cells **of** the primary meristem become differentiated, some becoming elongated and fusiform and others forming the spiral vessels. Trace them further and further from the growing point and notice that the differentiation constantly increases.

38. **On the sides** of the section, just behind the **conical** point, one or two elevations, the apices of the **axes of** lateral **buds of** the succeeding **season. Draw the** apex.

B. THE LEAVES. Cut **a transverse** section of one of the older needle leaves below the **middle.** Examine with a low power, and note

1. The *shape of the section.*

2. The three distinct *regions* it presents :

   a.　The narrow outer *cortical region*, whitish in color.

   b.　The central **oval** *fibro-vascular region*, bounded **by** a **distinct** chain of cells, the bundle sheath.

   c.　**Between** these two regions, a zone of green (greenish **even** in alcoholic specimens) parenchyma, the *mesophyll.*

3. The **number and position of** the *resin ducts.*

4. **Make** a sketch **of the** section.

Examine with a high power, and study

5. The *epidermal cells.*   Note

 a. The very thick *walls,* their *cavities* nearly or quite obliterated.   The outer layers of this thickening are cuticularized.

 b. The *cuticle,* quite thick and dipping as a thin wedge between the cells.

 c. The crack-like *pits* radiating from the cavity.

 d. The enlargement of the cell which forms the corner of the leaf.

 e. The *stomata.*   Study their structure carefully, noting

  i. The peculiar shape of the epidermal cells above the stoma, the outer wall, about as thick as the adjacent cells of the epidermis, prolonged upward to form a ridge overarching the *outer chamber* of the stoma.   Observe the cavity of these cells, much larger than those of adjacent cells.   At the bottom of the outer chamber,

  ii. The *guard cells,* their shape and the thickening of their outer walls.

  iii. The large intercellular space beneath the guard cells, the *inner chamber* of the stoma.

6. The usually single, in places double or triple, row of small cells underneath the epidermis, the **hypoderma.** Note

 a. The *shape,* and the thickness of *walls.*

 b. Where the greatest number of cell-rows occurs.

 c. The well-defined *middle lamella.*

 d. That the hypoderma is interrupted at each stoma.

7. Draw a stoma with a few of the adjacent epidermal and hypodermal cells.

8. The *mesophyll.* Note

   a. **The** *shape* of the cells, **and the** *number of rows* between the hypoderma and bundle **sheath.**

   b. The *infoldings* **of** the wall, dividing the cavity into recesses. Observe the position of the most prominent of these infoldings in the outermost **row of** mesophyll cells. Observe occasionally **(usually** near a stoma) branched cells. Determine the relation of these to the cells with simple infoldings.

   c. In fresh specimens, **the** abundant *chlorophyll.*

   d. The *resin ducts ;* compare their structure with those of the stem. Notice the thick walls of the **cells of** the sheath.

9. Draw a few mesophyll cells showing also **a resin duct.**

10. The *fibro-vascular region.* Study

    a. **The** *bundle sheath ;* **shape** and **contents of the** cells.

    b. The two masses of small cells, the *fibro-vascular bundles,* somewhat separated from each other and obliquely placed. Note the well-marked division into two areas :

       i. The *xylem,* **next the flat** side of the leaf, consisting **of spiral** and reticulated vessels and tracheides, arranged in radial rows.

       ii. The *phloem,* **next** the convex side of the leaf, consisting chiefly of undeveloped sieve cells.

       iii. The radial **rows** *of parenchyma* (like medullary rays), **passing** through **both** xylem and phloem.

       iv. In the xylem area, occasionally a poorly developed *resin duct.*

v.   Draw one of the bundles.

c.   Between the bundles and more or less encircling them, especially next the convex side of the leaf, *fibrous tissue* consisting of large thick-walled cells with small cavities.

d.   On the side of the bundle pair toward the flat side of the leaf, large *thin-walled, mostly empty cells.*

e.   Filling the remainder of the fibro-vascular region and entirely encircling the parts named, large *tracheides* resembling the preceding, but with more or less conspicuous contents, and walls marked with bordered pits. Compare the markings with those of the tracheides of the stem, studying both face and section views.

f.   Draw a few cells of each tissue named outside the bundles.

Cut a longitudinal section through the central part of the leaf.   Examine with a high power, and study

11. The *epidermis.*   Note

a.   The *shape of the cells.*   Unless the section be quite thin, the epidermis will appear as a continuously thickened border of the section.   The end walls of the cells are hard to make out, even in the best sections.

b.   The irregular *cavity*, and innumerable *pits* which perforate the thickening layers.

c.   If a number of sections be made, one or more will traverse a line of *stomata.*   Note the shape of the outer chamber, the shape of the guard cells, and of the intercellular space below.

d.   Draw a stoma and the adjacent cells.

12. Underlying the epidermis, the elongated sclerenchyma cells, the *hypoderma.*   In sections passing through a

line of stomata, note the absence of any hypoderma, **except** short cells between **the guard** cells of adjacent stomata.

13. Draw a few cells of epidermis and hypoderma.

14. The *mesophyll*. Note how loosely it **is** arranged, with **an** intercellular space between the rows of cells, enlarging under **each stoma**. Note also the shapes of the cells, **the** apparent absence of infoldings in this view, **and** the number of cells in each row between the bundle sheath and hypoderma. The determination of the latter point will need close inspection and careful focusing. The infoldings seen in transverse section are now seen as apparent partitions increasing the apparent number **of** cells above the actual. **In** places **none** of these false partitions occur, and the **real number of cells may be** easily noted. **Draw a few rows of meso-**phyll cells.

15. The *resin ducts ;* note the sheath cells, elongated **and** thick walled ; the secreting cells, with thin wavy walls and prominent nuclei. Draw.

16. The *fibro-vascular region.* The various tissues of this region appear as strips in this section.

    a. The *bundle sheath ;* **a row of** elongated cells next the mesophyll. **Draw.**

    b. The *tracheides,* **on** both sides of **the bundles ; note** the shape and markings of **the** cells.

    c. **Note the** change in shape where this tissue adjoins the fibrous tissue, the cells becoming much elongated. Draw, showing both forms.

    d. The *fibrous tissue ;* greatly elongated thick-walled fibers with tapering ends, **next** the tracheides. Draw.

e. Next the xylem, large thin-walled mostly **empty cells.**

f. The *phloem ;* consisting of **thick-set, very long** cells, **with** slightly **oblique ends, usually** crowded **with protoplasm and containing large** nuclei. **Draw.**

g. The *xylem ;* note

　　i. *Tracheides* like those of the stem but poorly developed, with few markings.

　　ii. *Spiral vessels* like those of the stem ; variously placed **with respect to** the tracheides.

Cut the thinnest possible slice from the surface **of an old leaf and then cut a thin section from the same place. Mount both with the outer surfaces upward and examine** with **a** high power. **Studying the first, the slice of** *epidermis,* note

17. The arrangement of the *stomata.*

18. The two kinds **of** *epidermal cells,* **those lying near and in a** line of stomata and those lying between **the lines** of stomata ; **observe** the shapes. **In the** former note **the ridge formed by the** upturned **edges of the six** cells **which bound the stoma.**[11] **If this cannot be** readily **made out,** treat the specimen **with potash and** observe **again in a** few minutes. Draw **a few cells of each.**

In the tangential **section** from **beneath the epidermis,** note

19. **The** cut ends of the mesophyll cells ; shape and arangement. Draw.

---

[11] The student should not mistake the peripheral border of the ridge for the outer wall of the six cells. The cells mentioned are quite large, the central ones extending from one stoma to another and the others usually half that distance.

20. The guard cells of the stomata also may usually be seen. Draw.

Cut a transverse section of the base of a young leaf, collected at flowering time.

21. Compare with the transverse section of an older leaf. Note the presence of protoplasm in almost all the tissues, completely filling the cells. Clear with potash. Compare carefully each tissue with the mature form, noticing particularly the lack of differentiation of the tissues, especially in the fibro-vascular region.

Mount one of the scales which enwrap a young leaf. Examine with a low power, and note

22. The shape and arrangement of the cells.

23. The fringe at the free end of the scale. Notice of what each hair of the fringe consists.

Cut a transverse section of these scales by cutting a transverse section at the base of a pair of young leaves. The sections of the scales will float off when the leaf section is placed in water. Note

24. The number of cell rows in thickness ; the shape of the cells and thickness of the walls of some of them.

25. A trace of a fibro-vascular bundle in the center of some of the thicker scales.

C. THE FLOWERS.

1. The *stamens*. Tease out a portion of the wall of an empty pollen sac. Examine with a high power, and note

    a. The shape of the cells.
    b. The beaded appearance of the walls.

    c.   Draw a few **cells.**

Place an entire male flower, from whose pollen sacs the pollen has all escaped, between pieces of pith and cut transverse sections of the cluster. Chance sections of the walls of the pollen sac will thus be obtained. Examine with a high power, and note

    d.   The number of cells in thickness.

    e.   The reticulated thickening of the lateral walls, which gives rise to the beaded appearance seen in the surface view. Draw.

Break open two or three pollen sacs and mount the *pollen.* Examine with a high power, and note in each grain

    f.   The *three lobes* into which it seems to be divided : a central one, the essential part of the grain ; attached to this two vesicular protrusions or *wings* with wrinkled surfaces.

    g.   In the *central lobe* make out

        i.   The *double wall* of the cell ; the outer part, the **extine,** rather thick and having its slightly roughened external portion expanded into the vesicular wings ; the inner, the intine, very thick and transparent.

        ii.   The *contents* (treat with iodine) ; protoplasm, abundant starch, and sometimes one or two clear-looking drops of oil.

        iii.   The *division* into two cells : one very large, containing the starch and oil ; the other very small, at the end of the central lobe furthest from the wings, best seen when the grain is lying on its side.

Treat the iodine-stained pollen grains just examined with 75% sulphuric acid. Press gently on the cover glass with the handle of a dissecting needle. Examine with a high power, and note

    h.   The empty *extine*, distorted by the pressure, and stained yellow.

    i.   The *intine*, blue, much swollen, and either empty or still containing the protoplasm, starch and oil. If empty the smaller cell can usually be well seen.

    j.   The yellow *protoplasm*, dark blue *starch grains* and clear *oil drops*, escaped from some of the pollen grains.

    k.   Draw a pollen grain, showing all its parts.

2. The *cone.* Take a cluster of female **flowers, bisect it** longitudinally, and from one of the halves cut longitudinal radial sections. **Treat** with potash. Examine with a low power, and selecting a section which has passed through an ovule, note

    a.   The *central axis* bearing the *bracts*, each subtending a *carpellary scale*, to whose upper surface the *ovule* is attached.

    b.   The body of the ovule, *nucellus*, surrounded by

    c.   The *integument*, which is prolonged beyond it.

    d.   The continuity of the nucellus and integument with the carpellary scale.

    e.   The orifice in the integument at its proximal end, the *micropyle*.

    f.   Draw, showing the above points.

Examine with a high power, and notice

    g.   That the cells of the nucellus, integument and scale are all alike **parenchymatous** and filled with protoplasm.

Dissect out a carpellary scale from the central part of a year-old cone and cut a series of longitudinal sections, including about the middle third of the ovule. Mount all the sections, and treat with potash. Examine with a low power, and note

   h.   The *parts of the ovule :* integument, nucellus and micropyle.

   i.   The *parts of the scale :* the scale proper and the wing of the seed. Notice that the tissues of the scale are continuous with those of the wing and ovule ; a faint trace of the coming lines of separation may however be detected.

   j.   The differentiation of the tissues of the scale into two kinds : the one of densely packed small cells forming an outer layer with deeper seated fibers ; the other of looser larger cells, forming the intermediate portion.

   k.   The differentiation of the integument of the ovule into two layers, the outer of densely packed small cells, the inner of looser larger cells.

   l.   The discoloration of the apex of the nucellus.

   m.   The presence of a large cavity in the nucellus, the *embryo-sac*, filled with a delicate transparent tissue, *endosperm*.

Examine with a high power, and note

   n.   That the body of the nucellus is almost entirely displaced by endosperm cells.

   o.   The wall of the embryo-sac ; wavy and usually broken away from the remaining cells of the nucellus in cutting the section.

   p.   The *endosperm cells ;* observe
      i.   The delicacy of the *walls*.
      ii.   The *contents ;* thready protoplasm and a very large round nucleus with a nucleolus.
      iii.   Draw a few endosperm cells.

q.   Near the outer end [12] of the embryo-sac, one or two much larger cells, the *archegonia* or **corpuscula.**[12] Observe the distinct **row of** endosperm cells, smaller than the others, which surrounds the archegonia.

r.   Occasionally one **or two** pollen grains having shed the **extine,** may **be** found in the micropyle, **and still more** rarely, some may be **found** which **have** begun to emit their tubes.

s.   **Make a** diagram **of** the ovule **and all its parts,** - together with the wing and carpellary scale.

ANNOTATIONS.

The Scotch pine raises a strong tall stem above **the** ground **for** the purpose of better exposing **its** leaves and fruits to **the air and** sunlight. This **habit is** correlated with the **excessive** development **of the** fibrovascular system, **which includes** all **the tissues of the** mature stem, with the exception of **a** trifling amount **at its center and** circumference.

Not only **is there** provision for continued growth in length by the formation **of** terminal buds, as in Adiantum, but there is also provision for growth in diameter. **A part of the** tissue, **from** which the fibro-vascular bundles **are formed, lying** between the xylem and phloem, **retains the power** of division and by annual increase **in the** number **of** cells, **chiefly in a radial** direction, the **thickness of the** bundle is increased. **The** difference in the size **and shape of the cells** added to the xylem in the spring **and** autumn gives rise to the

---

[12] *I. e.,* the end nearest the micropyle.

[13] Frequently not well developed at this time.

so-called annual or growth-rings which can be seen in the wood.

The scales which cover the stem, though called by the same name as the brown chaffy appendages to the stem of the fern, are not trichomes like them, but leaves. In addition to these scale leaves, which perform only slightly the function of true leaves, there are the needle leaves, upon which the foliage work chiefly depends. The delicate scales which enwrap the bases of the needle leaves are not trichomes, but leaves, as the rudimentary fibro-vascular bundle in them shows.

The different mode of arrangement of the scale leaves (and consequently of the dwarf branches) upon the terminal and lateral shoots is worthy of notice.

Concerning the homology of the parts of the male and female flowers, more especially the latter, there has been and still is much controversy. It is generally admitted that each cluster of stamens constitutes a single male flower. The scales which bear the pollen sacs on their under sides are homologous with leaves, as is shown by their position and anatomical characters and occasionally in teratological changes.[14] Moreover, the flower is subtended by a bract, and the floral axis bears several (usually three) bractlets below the stamens.[15]

As first announced by Robert Brown [16] the ovule in the pines and their allies is naked, i. e. it is not surrounded, as in the vast majority of flowering plants,

[14] Eichler, Blüthendiagramme, p. 59.

[15] Cf. Strasburger, Das botanische Practicum, p. 469.

[16] Appendix to Botany, Capt. King's Voyage, iv, p 103.

by an ovary; whence the entire group of plants having this character are called gymnosperms. Latterly, there has been much controversy **as to** the nature of the carpellary scale and whether the ovule is really or only apparently naked. The latter question involves the determination **of** the nature of the integument of the ovule. It is held on the one hand that the ovule consists of nothing but a nucellus, and that the coat surrounding this nucellus is the homologue **of the** wall of the ovary. On the other hand it is contended that this structure is the true integument of the ovule and that the scale which bears the ovule is an open carpel or pair of carpels.[17] In the laboratory directions we have adopted the latter view, calling the organ which bears the ovules a carpellary scale. This carpellary scale is theoretically "**com**posed of two leaves of an arrested and transformed branch from the axil of the bract, which **are in the** normal manner transverse **to** the subtending bract, * * * each bearing an ovule on its dorsal [as to position, upper] face; the two are coalescent into one by the union of their posterior edges, and the scale thus formed is thus developed with dorsal face presented to the axis of the cone, the ventral to the bract. It is therefore a compound open carpel composed of **two** carpophylls. **This** character of being fructiferous **on the back** or lower side of the leaf occurs in no other phænogamous plants."[18]

---

[17] References to extensive literature of this discussion may be found in Gray, Struct. Bot., p. 272. For a general statement of views and summing up of argument see Eichler, Sind die Coniferen gymnosperm oder nicht? Flora, 1873, p. 241. Consult also **Sachs**, Text-book, 2nd Eng. ed., footnote, p. 507. **From** references in **these** places the whole subject may be traced.

[18] Gray, Struct. Bot., p. 273, footnote.

As soon as the male flowers begin to scatter their pollen to the wind, the axis of the young cones elongates, separating the carpellary scales sufficiently to allow the pollen to be blown in between them, and to slide down, guided by the keel, to the prolongations of the integument. These prolongations subsequently roll inward, thus carrying any grains which may have become attached to them to the apex of the nucellus. After this process of pollination is accomplished the bracts cease to develop and likewise the now useless keel.[19]

The minute anatomy of the Scotch pine presents many points of considerable interest.

True tracheary tissue is formed only at the periphery of the pith, where a cluster of spiral, reticulated and pitted vessels occurs at the apex of each woody wedge.

The tissue of the wood is almost exclusively made up of tracheides, on whose radial walls are bordered pits. As these walls, originally thin and plain, increase in thickness irregularly, a part of the thickening on each side of the primary wall grows away from it to form the arched "border" of the small aperture which remains. For some time the primary wall remains as a membrane separating the two cells; when finally this is destroyed there is free communication between the contiguous cells. [20]

The thin delicate walls of the cambium allow great activity of the contained protoplasm, which results

---

[19] Strasburger, op. cit., p. 476.

[20] Cf. Strasburger. Bau und Wachsthum der Zellhäute, p. 43, taf. iii. For figures cf. Sachs, Text-book, p. 25.

in the formation by division of many new cells. The older cells on the axial side become gradually transformed into the tracheides and those on the peripheral side into the elements of the phloem.

Replacing the tracheides of the xylem are the sieve cells of the phloem. The radial walls of the larger cells have on them clusters of small perforations which are known as sieve plates or disks. These sieve plates are homologous with the bordered pits on the tracheides of the xylem.[21] At a little distance from the cambium they become covered with a homogeneous substance, the so-called callus plate, which completely interferes with the function of the sieve cells. Though this callus plate is subsequently dissolved, the sieve cells never regain their activity, the protoplasm having by this time disappeared from them.[22]

The cells with brown and crystalline contents are the true phloem parenchyma. A single row of them is formed each season, so that the age of the stem may be determined by these,[23] as also by the growth rings of the xylem.

The general arrangement of the tissues of the bundles is in contrast to that in the fern. The xylem and phloem here lie side by side, whence the bundle is known as collateral.[24]

The rigidity of the leaves of the pine is due to the thickening of the cells of the epidermis, together with the development of the layer or layers of hypodermal fibers.

---

[21] Strasburger, Das botanische Practicum, p. 143.
[22] Strasburger, op. cit., p. 147.
[23] Strasburger, op. cit., p. 146.
[24] Russow, Vergl. Untersuch., fide DeBary, Comp. Anat., p. 319.

Although the guard cells of the stomata appear at first sight to be deeper seated than the epidermis, observation teaches that they have been pushed down by the crowding over them of the adjacent epidermal cells, and here, as always, belong to the epidermis. This is confirmed by examining younger stomata.

The partial partitions by which the mesophyll cells are distinguished are explained by Sachs[25] as intrusive foldings due to local growth of the wall at the point where the fold occurs. Corry[26] however asserts that there is no real, but only apparent ingrowth, which is caused in this way: when the cells are still small their nuclei are attached to the protoplasm lining the wall by delicate protoplasmic strands one or more of which at a later period become converted into cellulose thus attaching the nuclei firmly to the wall. When the cell enlarges these points are firmly held near the nucleus. Since some of the strands soon break, many of the infoldings are shallow while others holding, cause deep infolding.[27] The purpose of these infoldings is considered by Haberlandt to be to secure a greater surface on which to display the chlorophyll bodies. Corry says of them: "They perform at all events a very obvious and noteworthy function in forming the intercellular spaces beneath the stomata in Pinus, and in producing air channels

---

[25] Text-book, 2nd Eng. ed., p. 74.

[26] On some points in the structure and development of the leaves of Pinus sylvestris.—Proc. Camb. Phil. Soc., iv (1883), p. 344 et seq.

[27] Similar infoldings in leaves of *Elymus Canadensis* and other grasses are described by Kareltschikoff (Bull. Imp. Soc. Nat. Moscow, xli [1868], p. 180) and in *Caltha palustris*, *Anemone nemorosa* and a number of other plants by Haberlandt (Oester. Bot. Zeit., xxx [1880], p. 305).

between the cells forming the several rows of palisade tissue." [28]

The four bundles of each pair of leaves **have the** normal orientation, the xylem portions **all** facing a common center and the phloem the periphery. The imbedding of the bundles in a mass of colorless tissue surrounded by **a** sheath is common among the pines and their allies.

In this central tissue many of the cells are tracheides (see fig. 7), as pointed out in the laboratory part ; they are arranged in a special manner and are characteristic of *Coniferæ.* [29] These tracheides during the activity of the leaf contain water, [30] and hence have been called transfusion tissue by H. v. Mohl[31] and others.

The existence of occasional poorly developed resin passages in the xylem of the leaf bundles is to be noted, as it has been denied by Corry[32] and **Van** Tieghem.[33] (See fig. 7 **r**).

In comparing the reproduction of the pine with that of the fern and earlier forms we find advances of **much** interest. In the fern, as in the moss and liver-**wort,** the spore grows into a structure, which bears the reproductive organs. In the moss and liver-wort this sexual or thalloid stage comprises by far the larger part of the life cycle, while the asexual stage (the so-called fruit) is small and quite unable to lead an independent existence. In the fern **the** thalloid

---

[28] Op. cit., p. 355.
[29] Cf. DeBary, Comp. Anat., p. 378 et seq.
[30] Strasburger, op. cit., p 234.
[31] Bot. Zeitung, 1871, No. 1, 2.
[32] Op. cit., p. 359.
[33] Ann. Sci. Nat., Sér. V, xvi (1872), p. 189.

stage is much reduced, although still green and able to maintain itself for a limited time, while the asexual stage is the conspicuous part of the plant, in fact the only part usually noticed, except by students and fern propagators.

From the fern to the **pine** is too great a step to be **well** understood without considering some intermediate **type.** **Some species of** Selaginella would answer this **purpose admirably, and it** is to be regretted that no **species** is sufficiently common in this country, either wild or cultivated, to permit the introduction of direc- **tions for** its study in this manual. It must therefore suffice to mention one feature of Selaginella **indispens- able** for a clear understanding of the **subject in hand.**

Selaginella, instead of having only one sort of spores, as in the ferns and liverworts, has two, one small (micro- spores), the other large (macrospores). When these **spores** vegetate, the prothallium from the smaller one **bears the male** organs (antheridia), and that from the larger the female organs (archegonia). A very marked **feature is that the prothallia are** greatly reduced, so **much so in fact** that they never leave the spore or become green, and the one **from the smaller spore is** even reduced to a single small cell.[34]

To return **to** pine, we shall find that the reduction of the sexual stage **or** prothallium is carried a step, and quite a long step further than in Selaginella, while the asexual stage is augmented in the **same** proportion. **The latter in** fact is the pine tree—the whole plant one would naturally say. It must be borne in mind that in

---

[34] For further description see Bessey, Botany, p. 385 ; Sachs, Text- book, p. 468.

the fern **the asexual** plant produces spores, and that in
Selaginella, a more advanced type, it does also, but **of**
two sorts. Does the pine likewise produce **spores?**
Certainly, although we have so long called them pollen,
that we are inclined to forget **their** true relation,
which would be better indicated by the term pollen
spores, used **by** DeBary."[25] These pollen spores cor-
respond to the microspores **of** Selaginella, and like
them have the prothallium reduced **to** one or a few
cells, **but** unlike them **do** not produce antherozoids.
This, however, is a matter of adaptation. **Wherever**
there is water to transport the fertilizing element
from the male to the female organs, it is usually
an active body (antherozoid), as in Adiantum, Atrichum and Marchantia, with an exception **in** Spirogyra,
while if it must be transported through the air or the **inte-**
rior of plant tissues a tube leads from the antheridium
to the archegonium as in Microsphæra and Cystopus.
Pine like other flowering plants has the spores carried
bodily through the air in order to bring them into
proximity to the female element, then a tube (pollen
tube) develops, which connects the male and female
organs. Turning now to the female part, which corresponds to the macrospore of Selaginella, it (now called
the embryo-sac) is found **so** greatly reduced **that it**
never leaves the **place in** the mother plant where
formed. The prothallium **is** represented by the primary endosperm. The archegonia themselves are
much simplified as might be expected. They arise
from superficial cells of **the** endosperm (prothallium).

---

[25] Morph. **u.** Biolog. **d. Pilze, Mycet.** u. Bacterien, 1884, p. 140.

Within each is a large nucleated germ cell or oosphere, the part to be fertilized.

The process of fertilization is as follows: The pollen grains having been lodged in the micropyle upon the apex of the nucellus, the extine is burst and slipped off by the swelling of the intine and its contents. By a local growth the intine extends into a tube into which the contents of the larger cell pass by a streaming movement, the smaller cell remaining inert. This pollen tube pushes its way slowly between the cells of the nucellus until it reaches the germ cell in the embryo-sac. Shortly afterward a nucleus almost as large as that of the germ cell appears below the end of the pollen tube. It is to be supposed that it has passed through the wall of the tube, and it is to be regarded as homologous with the body of an antherozoid. The two nuclei fuse into one, which passes to the end of the germ cell opposite the neck where it gives rise to several four-celled layers, one above another, the lower four of which form the beginning of the embryo."[36] This process of fertilization requires in Pinus sylvestris a little more than a year between the beginning of the growth of the pollen tube and the consummation.

The fertilized germ cell grows at once into the young plantlet (embryo), as in the fern, but at this stage, unlike the fern, it stops for awhile, and in the passive, well protected condition of a seed may pass a long period before it resumes its growth. This, again, is a special adaptation. All the plants heretofore considered are fully equipped for the dispersion of each

[36] Cf. Strasburger, op. cit., p. 481 et seq.

succeeding generation through their sexual or asexual spores, or the division of the vegetative members. In the pine the young plantlet is developed before leaving the parent, and were it to continue to grow would either live wholly upon the parent, or be brought into such close competition with it, that the species would speedily become extinct. Therefore, to provide for the proper dispersion of the offspring, the young plantlet is suitably protected, and provided with food for its first growth when again resuming its development, separated from the parent, and wafted away by the wind in the utmost security.

This is one of the most characteristic features of the higher plants, from which they might better have been named seed-bearing plants, than flowering plants.

It is also worthy of notice that the primary endosperm which is formed during the first year of the fruit, and on which the archegonia arise, is subsequently destroyed by the deliquescence of the cell-walls; and from the protoplasm thus set free there is produced in the spring of the second year what may be called secondary endosperm which, with the growing embryo, fills up the embryo-sac and displaces the most of the tissue of the nucellus."

---

[31] Cf. Sachs, Text-book, 2nd Eng. ed. p. 521.

FIELD OATS.

Avena sativa L.

PRELIMINARY.

THE cultivated grass known as oats is too familiar to need description. Specimens should be collected at the time when some flowers of the panicle are expanded and others are yet in the bud. This plant begins to bloom shortly after the panicle is liberated from the sheath. The time of blossoming is so little marked by external changes that there is great danger that specimens will be collected too late. Care should be taken in lifting the plants from the ground not to detach the empty grain from which it grew, which will almost certainly be done if the plants are pulled up. They should be dug and the dirt shaken gently from the roots, which may be further cleaned by washing.

The requisites for the complete study of the plant are entire plants, preserved in alcohol; a handful of threshed oats; alcohol; magenta; potassic hydrate; and iodine.

LABORATORY WORK.

GROSS ANATOMY.

A. GENERAL CHARACTERS. Note the four parts of the plant :

1. The *roots*.

2. The **upright main axis, the** *stem*, **with** numerous branches near the top.

3. The lateral appendages of the stem, the *leaves.*

4. The surface appendages on the roots and leaves, the *trichomes,* in both instances extremely minute.

B. **THE** ROOTS. In a plant which **has** the emptied grain from which it grew still attached, **note**

1. The small group of roots arising from one end of the grain, the strongest of which is the *primary root.*

2. The stem emerging from the other end, the *first inter-node* of the stem.

3. **At** a certain point,[1] the *second node* of the stem, **a whorl of** *secondary roots.*

4. **At** one or **two** succeeding nodes, a like whorl **of sec-**ondary roots.

5. Make a diagram, showing the position of the roots and their relation to the lower part **of** the stem.

Cut a transverse section **of** one of the large secondary **roots.** Examine by transmitted light. Note

6. The round central spot **of** firmer tissue, the *fibro-vas-cular bundle.* The openings in it are the **larger** *vessels.*

7. **The loose, pith-like** *cortical portion.*

8. The *root-hairs,* attached **to** the edge.

9. Draw.

---

[1] Known to agriculturists as the "tillering point." The length of **this** first internode depends to a considerable extent on the depth of planting **the** seed.

Strip off the cortical portion of one of the large secondary roots.   Notice

10. The slender, strong fibro-vascular axis which remains.

Examine some plants three or four days old, which have been grown on the surface of wet blotting paper.   Note

11. The position of roots and stem with respect to the grain.

12. The abundant root-hairs.   Notice their relative length on different parts of the root, and where absent.

13. The opaque tip of the root covered by the conical *root-cap*.

C. THE STEM.   Notice that it is completely encased by the sheathing bases of the leaves.   Uncover a portion of the stem by removing one of the leaves and its sheath, and note

1. Its *shape*, and polished *surface*.

2. Its *nodes* and *internodes*.   Bisect the stem longitudinally through a node and a portion of an internode.   Note

   a.   The solid node forming a partition between the cavities of the internodes.

   b.   Draw.

Look through the split stem at a bright light, and note

3. The numerous threads, traversing the stem lengthwise, the *fibro-vascular bundles*.

Cut a transverse section and examine by transmitted light, and note

4. A very firm, more opaque external layer, the *cortical layer*.   Notice its variable thickness.

5. In the cortical layer, pairs of darker spots.   These are clusters of *chlorophyll-bearing cells*.

6. The remainder of the section made up of large rounded cells, *parenchyma*, scattered through which **are**

7. Masses of firmer tissue, the fibro-vascular bundles, each having three or four openings, the *vessels.*

8. Draw the section.

Cut a number **of** longitudinal sections ; in them **make out**

9. The denser cortical portion.

10. The more transparent **parenchyma.**

11. The fibro-vascular bundles.

12. In a section not passing through a fibro-vascular bundle, the strip of darker chlorophyll-bearing tissue under a **very** narrow cortical layer.

13. **Draw a section, showing as** much **as possible of the** structure.

D. THE LEAF. Note

1. Its sheathing *base.* Observe the extent of stem covered **by** each sheath.

2. The split in the sheath ; its position and extent.

3. **On the upper surface at the** point **where the** sheath **ceases,** a thin membranous outgrowth, **the ligule. Notice** its shape and apex. **Draw.**

4. **The place** of attachment of the leaves.

5. The remainder **of the leaf,** the *blade.* **Note**
   a. Its *shape.*
   b. The numerous *veins ;* their direction and relation to the ridges.
   c. The green tissue (bleached by alcohol) between the veins, the *mesophyll.*

Cut a transverse section of the blade, and note

6. The variable *thickness* of the leaf.

7. The sections of the fibro-vascular bundles.

8. On the upper edge, large cells between the ridges, the **hygroscopic cells,** which cause the leaf to roll when dry.

E. THE FLOWERS.

1. The arrangement of the flowers, *anthotaxy.* Note

    a. The central stem of the flower cluster, the *main axis of inflorescence.*

    b. Its lateral branches, *secondary axes.* Notice their relative lengths.

    c. That some of the secondary axes are branched, others not, thus constituting a **panicle.**

    d. Make a diagram of the mode of branching.

    e. That each ultimate branch bears **not** a single flower, **but a cluster of three** (sometimes two) flowers, **a spikelet,** at the thickened extremity. The **entire** inflorescence **is thus** compound, **a** *panicle of spikelets.*

    Detach a spikelet, **and note**

    f. **Two bracts at the base of** the spikelet, completely inclosing the flowers, the **empty** glumes. Notice the position of **these** glumes with respect **to** each **other** and their points of attachment. **Detach them,** and note

        i.   Their *shape.*

        ii. The parallel veins, *nerves ;* **the** number in each glume, the termination **above,** the delicate cross (anastomosing) **veinlets.**

        iii. Draw.

g. The three flowers inclosed by the empty glumes ; **their** relative size[2] and position on

h. The flattened **axis** on which they **are borne, the** *rhachis* of the spikelet.

i. The *tufts of minute hairs* at the base **of the lowest** flower.

j. Draw a spikelet, showing the empty glumes and flowers separated from one another.

2. **The** *structure of the flowers.* Detach the lowest flower **in the** spikelet. Note

    a. **The** bract, **flowering glume,**[3] which almost incloses the flower. It sometimes bears a long bristle-like appendage or **awn on** its outer surface ; note position when present. Detach this bract entire, and note the size, shape, surface, texture, notched apex and number of nerves. Draw.

From another **flower** cautiously **detach the** flowering glume by cutting and tearing **it** away piece **by piece,** leaving only a bit of its base, being careful **not to** injure

    b. The *flower proper.* Observe

        i. A large bract-like body, the **palet,**[4] its infolded margins, shape, nerves, and the presence and position of the trichomes on its outer surface ; **contrast it with the** flowering glume. Draw ; also make **a diagram of a** transverse section **at its** middle.

        ii. **Two small** bract-like bodies, **the lodicules,** situated between the edges of the palet. **Ob**serve their shape and texture. **Draw.**

---

[2] The third is rudimentary and lies close to the inner side **of the upper** flower.

[3] Called the lower palet in most systematic works.

[4] Called the upper palet in most systematic works.

iii. The three similar *stamens*. Examine one carefully, and note three parts :

α.   The slender thread, filament, carrying at
its apex

β.   A two-lobed body, the **anther**; note
the deep groove lengthwise of each lobe,
and the point of attachment of the filament.

γ.   Tear open an anther, or examine one
which has burst, and notice the cavities
containing *pollen ;* the color and powderi-
ness of the grains.

δ.   Draw a stamen.

iv. The hairy body in the midst of the stamens,
the pistil.   Note its three parts:

α.   The large, top-shaped part at the base,
clothed with white hairs, the ovary.

β.   The two thread-like bodies arising from
the top of the ovary, the **styles**.

γ.   The numerous branches[5] of the styles
arranged like the barbs of a feather, the
stigmas.

δ.   Draw a pistil.

Cut a pistil in two longitudinally between the styles,
and notice

ε.   The thick but delicate *wall of the ovary*.

ζ.   The *ovule* of denser tissue closely adhering to it, and mostly occupied by

η.   A cavity, filled when growing with the
transparent *endosperm*, which cannot now
be easily detected.   In this cavity notice

---

[5] If hidden by many adherent pollen grains, brush them off with a camel's-
hair brush.

Θ. The early stage of the young plantlet of the seed, the *embryo*.

ι. Illustrate **with** diagram.

Compare with this flower the second and third **flowers of** the spikelet. Note, in the latter, the absence of the inner organs, leaving only the flowering glume, palet, and some-times the stamens.

F. THE FRUIT. Study ripe oats which have **been threshed or** shelled out in the hand. Strip off the chaff (flowering glume and palet) which incloses the fruit. Note

1. The white *hairs* which cover it, especially at the upper end.

2. The longitudinal *groove ;* its position as to the palet.

3. The *scar* at the base of the grain opposite **the** groove, marking the position of the plantlet **within.**

   Cut across the middle of a grain, and note

4. The depth **of** the groove, **and** the uniform floury *contents ;* test with iodine.

5. That the wall of the ovary and the coats of the ovule have become so closely united and thin as to be indis-tinguishable, thus constituting the fruit a **caryopsis** or *grain.*

6. Draw the section.

From a **soaked** grain carefully remove the "skin" (the wall of the ovary **together with** the seed coats) on the side opposite the groove, **from the** lower **end to** the **middle.** There will then be seen a **face** view of

7. The *embryo.* **Note**

   a. The large elongated-triangular **body** forming the

upper part of the embryo, the *cotyledon* or scutellum.

b.　　The pointed lower extremity, the *root sheath.*

c.　　Near (below) the center of this face of the embryo, a minute bud, the *plumule.*

d.　　Just below the base of the plumule, a very short stem, the *caulicle.*[6]

e.　　Draw the embryo as it lies exposed.

Bisect a grain longitudinally through the groove.　Mount also a thin section from the cut surface.　Note

f.　　The scutellum, with its back against the starchy part of the seed, its face just under the "skin" at the upper part of the embryo.

g.　　The plumule, on the face of the scutellum, at the upper end of

h.　　The caulicle; easily recognized as the whitish part where the scutellum and plumule merge. At its lower extremity is

i.　　The root, a small rounded point, over which is

j.　　The root-sheath, which forms the lower extremity of the embryo.

k.　　Draw the section.

Take a series of transverse sections from the bottom of the grain upward.　Examine the successive cut surfaces and, comparing with the longitudinal section, determine the various parts seen, root-sheath, root, caulicle, plumule, scutellum.　Draw those which show the section of root and root-sheath, and the section of plumule and scutellum.

MINUTE ANATOMY.

A.　THE ROOTS.　Cut a transverse section of one of the lateral roots at a little distance from the stem.　Examine with a low power, and note the two regions :

---

[6] Ill defined and difficult to see.

1. The cortical, thin-walled cells.

2. The fibro-vascular, thick-walled cells.

Examine with a high power, and note

3. The thin-walled *epidermis.* Observe **its** irregularity, and the mode of attachment of the root-hairs. Draw.

4. The *cortical parenchyma,* with sclerenchyma either **intermixed,** or in older roots forming an outside layer. **Draw.**

5. The *bundle sheath ;* the relative thickness of outer and inner walls ; the pits. Draw.

6. The *fibro-vascular bundle.* Study
   - *a.* The smaller thick-walled cells constituting most of the bundle.
   - *b.* The vascular tissue ; four to **six** (sometimes **more)** large *pitted vessels* symmetrically disposed. **Between** each **of** these **and** the bundle **sheath (also some-** times **near the center)** a dozen (more **or less) of** smaller pitted **vessels.**
   - *c.* Numerous channel-pits in all the thick-walled cells.
   - *d.* Draw a portion of the bundle.

Cut a longitudinal section of the **same root.** Examine with a **high** power, and note

7. The epidermal cells. Observe **the** bases of **root-hairs,** and their relations to the epidermal cells. **Draw.**

8. Elongated **cortical** parenchyma and occasionally sclerenchyma. **Notice** the pits. Draw.

9. The bundle sheath ; cells elongated, rather difficult **to** distinguish. **Draw.**

10. **The** fibro-vascular **bundle.** Study

> *a.*   The **pitted** fibrous cells, *tracheides*, which constitute most of the bundle.
>
> *b.*   The one or two *pitted vessels*.
>
> *c.*   Draw, showing both vessels and tracheides.

Mount about one centimeter of the tip of a root from plants that have been grown upon blotting paper. Examine with a low power, and note

11. The *root-hairs*. Observe their relative length.

12. The *root-cap ;* the outer cells sloughing off. Draw.

Study the root-hairs with a high power. Notice

13. The **shape**, mode of attachment and contents. Draw.

Cut a median longitudinal section of the tip of a root, including the **root-cap.**[1] Treat with potash, examine with a high power, and note

14. The blunt, or **even notched** tip of the root proper.

15. The sharp conical root-cap. Note the shape of the **cells** near the root-tip, and the changed shape near the periphery.

16. The *growing point*, a cluster of small cells, just back of the root-cap, in the middle of the root-tip.

17. A short distance behind the tip of the root, the slight differentiation of the tissues into three regions :

> *a.*   A central one, the **plerome.**
>
> *b.*   An outer one, consisting of a single row of cells, the **dermatogen.**
>
> *c.*   Between the plerome and dermatogen, the **periblem.**
>
> *d.*   Trace these three regions down to the growing point, and notice their relations there.

---

[1] This is very difficult to do if fresh roots are used, but easier by using **roots** that have been kept for a few hours in alcohol. The student should cut a **series** of sections through the whole root. The **median** one can then be selected.

    c.   **Trace** them backward ; notice that the plerome **becomes the** fibro-vascular bundle ; the periblem, **the** cortical parenchyma ; the dermatogen, **the** epidermis.

Cut a transverse section of the oldest part of a root which has grown on blotting paper. Examine with a high power. Compare with the section of the large lateral roots, already studied. Notice

18. The origin **of the** root-hairs.

19. The differences in the fibro-vascular bundle, particularly the presence of **a** large axial vessel.

B. THE STEM. Cut a transverse section from one of the younger parts of the stem, *e. g.*, between the flower cluster and the first leaf. Examine with a low power, and note

1. An outer cortical **part,** of varying thickness, composed **of** small **dense-looking** cells, the *epidermis* and *hypoderma.*

2. In the cortex lighter spots, in pairs, at almost regular intervals, *chlorophyll-bearing parenchyma.*

3. An inner part, consisting of large empty parenchyma cells, the *fundamental parenchyma*, with fibro-vascular bundles at regular intervals.

    Examine with a high power. **Study**

4. **The *epidermis*. Note** the thick **walls, showing two layers, and** the cuticle. **Draw.**

    a.   **Observe in** some sections a pair **of** smaller, peculiar **cells in the** epidermis over an intercellular space **in the** chlorophyll-bearing parenchyma, the *guard cells* of a stoma. **The** two adjacent epidermal cells are also modified somewhat. **Draw.**

5. The *hypoderma.* Note the thick **walls of** the **cells**

which increase in size toward the parenchyma, but do not merge into it.   Draw.

6. The *chlorophyll-bearing parenchyma*.   **Note**

    a.   The *shape, size* and *arrangement* of the cells.

    b.   The thin *walls*.

    c.   The *contents ;* protoplasm and chlorophyll bodies which are green, if fresh stems are used.   Notice the position of the chlorophyll bodies.[8]

    d.   Draw a few cells.

7. The *fundamental parenchyma*.   Note the size and shape of the cells, and the triangular intercellular spaces. Draw a few cells.

8. The *fibro-vascular bundles*.   Notice the two series of bundles : the larger ones nearer the central cavity of the stem ; the smaller between the paired groups of chlorophyll-bearing parenchyma.   In the larger observe

    a.   The *external* **sheath,** an irregular layer of cells, with slightly thickened walls,[9] surrounding the bundle, and thicker on its peripheral side. Examine it in a section from an older part of the stem ; note the thickness of the walls.

    b.   The *tracheary tissue ;* on the right and left of the bundle two large *pitted vessels ;* toward the axial side one or two *annular vessels ;* between the large pitted vessels a transverse band of smaller pitted vessels.[10]

    c.   Between the annular vessels and the external

---

[8] They may be made plainer by staining with magenta.

[9] If it can not be discerned, stain slightly with magenta ; these cells take a deeper red than the rest.

[10] Stained a deeper red in the magenta-treated section.

sheath sometimes an *intercellular cavity* formed
by breaking in growth.

d.　Toward the peripheral side of the bundle a group
of thin-walled *conducting cells.*[11]

e.　Draw the bundle.

f.　Compare the structure of the smaller bundles **with**
the foregoing, noting differences.

Cut a number of longitudinal sections of the stem, and
examine with a high power.　Study

9.　The *epidermis.*　Note

a.　The thickened outer *wall ;* elongated shape; channel pits.

b.　The alternately long and short cells in **some**
sections.

c.　**Draw.**

Some of the sections will be likely to pass through a
stoma.　Examine

d.　The *guard cells ;* **note** the enlarged ends **and nar-**
row body.　Draw.

10.　**The** *hypoderma ;* note the extreme elongation and
tapering ends of the cells.　Draw.

11.　The *chlorophyll-bearing cells ;* **note their** shape, arrangement and contents.　Draw.

12.　The *fundamental parenchyma ;* **note the size and**
shape of the cells, **and** the thin places in **the walls.**
Draw a few cells.

13.　The *fibro-vascular bundles ;* note **in the** various **sec-**
tions [12]

a.　The slightly thickened, sparsely pitted, elongated

---

[11] Unstained with magenta.

[12] No one section can be found to show all points.

cells of the *external sheath* having slightly oblique end walls.

b. The delicate walls and elongation of the *conducting cells.*

c. The *pitted vessels*, large and small.

d. The *annular vessels.* Notice the various positions of the rings. Study their cut ends where the razor has passed along a vessel.

e. Draw a few cells of each tissue.

Cut a thin slice from the surface of a stem, examine with a high power, and note

14. The *epidermis.*

a. The cells above the hypoderma; shape and arrangement.

b. The cells above the chlorophyll tissue, including the stoma; shape and arrangement.

c. The numerous pits in the surface wall, and in the side walls beneath.

d. Draw.

C. THE LEAF. Cut a transverse section, and examine with a high power. Study

1. The *epidermis.* Notice

a. Its cuticularized outer *wall* with minutely uneven free surface.

b. The *guard cells.* Note
  i. The different appearance of these cells, according as the section has passed through the bodies or ends.
  ii. The small size and thick walls of the body, the larger size and thinner walls of the ends.

c. The *modified epidermal cells* adjoining the guard cells.

d. **Draw** various sections **of stomata,** with adjoining cells of the epidermis.

e. The modified large epidermal cells in the depressions on the **upper** surface, the *hygroscopic cells.* Draw.

f. The modified epidermal cells at the summit of each ridge ; sometimes teeth may be seen. Draw.

2. The *hypoderma.* **Note its** position and the character **of the cells.**

3. The *mesophyll,* all **the** chlorophyll-bearing **part of the** leaf. **Note**

a. The slight elongation of those cells next the epidermis, forming **palisade parenchyma.**

b. The large *intercellular space* under each stoma, and the numerous smaller ones in other places.

c. The abundant *chlorophyll bodies.*

4. The *fundamental tissue ;* often reduced **to** only one **row** of large empty cells surrounding the bundles.

5. The *fibro-vascular bundles ;* compare those forming the midrib and main veins of the leaf with those studied in **the** stem. Compare with these **the** bundles of the smaller veins, noting what **tissues are absent** from **them.**

6. Draw **a** portion of the section, including **a** large fibro-**vascular** bundle, and some cells of **the** mesophyll **and** fundamental tissue.

7. Make **a** diagram **of** the leaf section to show relative position and size **of the** different parts.

Strip off two pieces of the epidermis. Mount one piece **with the** outer surface uppermost, and the other with the **inner** surface uppermost. **Note**

8. The *epidermal cells*.

   a.   The *shape* of those lying above a vein, together with the short strong trichomes, each bearing a very sharp point, directed forward.

   b.   The *shape* of those lying among the stomata.

   c.   The *stomata*. Note

      i.   The regular *arrangement* in double or triple rows.

      ii.   The pair of narrow epidermal cells, which stand one on each side of the guard cells.

      iii.   The shape of the *guard cells ;* the thick walls of the body and thin walls of the ends.

      iv.   Draw, showing the several sorts of epidermal cells.

9. The shape and contents of the *mesophyll cells,* some of which will almost invariably adhere to the epidermis when stripped off. Draw.

10. Make a transverse section of the *leaf sheath,* and note its intermediate character between that of the stem and of the leaf blade already studied. Draw sufficient to show the various tissues, and their arrangement.

D. THE FLOWER.

1. The *glumes* and *palets.* Make a transverse section through the upper part of a spikelet and transfer it to the slide without disarranging the parts. Note

   a.   The thin-walled cells forming the inner portion, and the thick-walled cells forming the outer portion of each part. Draw from two or more regions.

   b.   The angles of the palets, bearing stiff trichomes. Draw.

2. **The *anthers*.** Tear off **bits of** the wall of an empty
   anther. Mount one outside up and the other inside **up.**
   Focus on the surface of the first, and note

   a. The *epidermis ;* its wrinkled *walls ;* the **shape of**
       its *cells.* Draw.

   Focus on the surface of the second, and note

   b. The *endothecium,* **the** layer of cells lining the
       anther. Observe

       i. **The** infolded *thickenings* **of the** side walls of
           the cells.

       ii. The *shape* of the cells.

       iii. Draw.

Cut a transverse section through the lower part **of a**
spikelet which has not bloomed, and transverse sections **of**
the anthers will be obtained. Notice

   c. The large inflated epidermal cells.

   d. The very narrow endothecial cells, with the thick-
       enings of the walls extending the full height, mak-
       ing it difficult to distinguish their outline.

   e. Draw a few of the two kinds of cells.

   Under low power, notice

   f. **The** two lobes of the anther, *thecæ.*

   g. **The** *connective* which joins them, containing a fibro-
       vascular bundle.

   h. **The four** *cavities,* appearing like **two after** dehis-
       cence. Usually the manner of dehiscence can
       be detected.

Using the same section, under high power, notice

3. The *pollen.*

   a. The *shape* of the cells.

   b. The small globular *protuberance* sometimes seen
       when the spore lies properly.

c.   The *optical section* of the wall ; its continuity interrupted at the protuberance.

d.   The *contents*.   Burst some **spores by** pressing
lightly on the cover-glass with a needle.   Note

   i.   Here and there entirely empty bursted sacs,
the *extine*.   Notice the minute roughening
of the surface ; the thin spot or opening,
through which in some cases when unburst the
intine protrudes.

   ii.   The contents of some spores surrounded by the
intine, escaped from the extine and become
much larger.   In some cases the protuberance
may still be seen.

   iii.   The *contents* of other spores free in the water
of the slide, showing innumerable fine granules.   Note their shape, and treat with iodine
to determine their nature.

e.   Draw an uninjured spore, showing its structure.

4.   The *styles* and *stigmas*.   Cut off one of the styles near
its attachment.   Mount and examine with a low
power.   Note

a.   The tapering style with

b.   Numerous undivided branches, the stigmas,
roughened with innumerable points.

c.   The grains adhering to the stigmas.

Examine with a high power.   Observe

d.   The thin-walled nucleated cells, forming the stigmas ; the proximal ends are overlapped by other
cells.

e.   The adherent grains.   Notice that some of the
spores have emitted through the perforation in
the extine a slender tube which penetrates the
stigma.   Notice that the granules of the pollen

spore also enter **this** tube. Observe that some spores have become empty.

f. Draw, showing structure **of** stigmas **and** the entrance of a pollen tube.

5. The *trichomes of the ovary.* Cut **off,** mount, **and** examine with a high power some of the trichomes which clothe the apical portion of the ovary. Note shape and contents. Draw.

E. THE FRUIT. Remove the chaff from a grain, **and** cut a transverse section near the middle, having previously soaked it in warm (not hot) **water** for **a** few minutes.[13] Note

1. While mounting, the abundant whitish powder which **escapes** into the water, clouding it more or **less.**

Examine with a high power **and note**

2. The outermost coat of the fruit, the *ovary wall,* some-times splitting into two layers ; the cells **can** only be made out with great difficulty.

3. **The** layer of large cells, containing granular proteid matters, chiefly *gluten.* Note shape, and test contents with iodine. Draw.

4. The large cells packed with granules of *starch,* made blue by the iodine. The outline of these cells is best **seen** when the starch has escaped from them.

5. The tip of the *embryo* **will** usually appear at **one side of the section.**

Cut a median longitudinal section **through the** groove of **a** soaked fruit. Treat **with potash** to clear up the embryo, and examine with a low power. Note

---

[13] An immersion of an hour or longer in cold water will answer the same purpose.

6. The three parts of the fruit : the walls of the ovary and gluten-containing cells ; the starchy part of the grain ; the embryo.

Study the *embryo* ; note

7. The long leaf, *scutellum*, next the starch.

8. The bud, *plumule*, near the base of the scutellum, showing one or two leaves.

9. The *root* near the base of the embryo, with its *root-cap*, and enveloped by

10. The *root-sheath* ; notice that it is continuous with

11. The short stem, *caulicle*, to which the scutellum is attached, bearing the plumule at its upper and the root at its lower end.

Examine with high power. Note

12. The tissues of the fruit, essentially as in the transverse section.

13. The tissue of the embryo ; parenchyma with much protoplasm.

ANNOTATIONS.

The division of the slender, slowly tapering stem of Avena into ring-like nodes and elongated internodes shows these features distinctly marked for the first time. The disposition of the material in the form of a hollow cylinder gives greater rigidity than would the same amount of material in a solid stem.

At some of the lower nodes of the stem the endogenous formation [14] of roots can be well seen, as young

---

[14] Cf. Prantl and Vines, Text-book, p. 22.

roots can be frequently found just breaking through the superficial tissues.

The leaves of oats are sharply distinguished into a sheathing base and a spreading blade. The membranous outgrowth, the ligule,[15] which is found at their junction, is common in leaves of this character.

The flower of oats, like that of the pine, is a metamorphosed shoot, in which the axis is the stem, and the lateral organs which it bears, leaves. At the base of each spikelet are to be found two glumes or bracts, which thus subtend and more or less completely inclose the whole cluster. At the base of each flower is a single bract, the flowering glume, having the flower in its axil. Concerning the homology of the palet and lodicules much discussion has arisen. Payer[16] asserts that the palet is a double organ and that the two keels on the palet are primitively distinct. Schacht[17] sees in the palet two parts of a trimerous whorl, of which the anterior part is suppressed. Röper, Wigand, Nägeli and others conclude from comparative and developmental researches that the palet is primitively single and takes on its two-keeled condition subsequently.[18] Hackel[19] believes "that the palea and the pair of lodicules (when two only) are each single, more or less bifid organs, and that they and the third lodicule, when present, must be regarded as two or three

---

[15] For a statement of its homology see Gray, Struct. Botany, pp. 106, 211.

[16] Organogénie de la fleur, p. 701.

[17] Das Mikroskop, 2 Aufl., p. 170.

[18] Cf. Eichler, Blüthendiagramme, p. 120.

[19] Untersuchungen über die Lodiculæ der Gräser, Engler's Bot. Jahrbücher, i, p. 336.

bracteoles inserted fore and aft on the floral axis below the flower, and he has made out a good case in favor of his view but perhaps not an unanswerable one." [20] Bentham [21] adds : " The search for homologies to the palea and lodicules in the orders nearly allied to the *Gramineæ* has met with but little success ; " and again, " The palea and lodicules of *Gramineæ* may represent perianth segments of an outer and inner series, though I by no means pretend to assert it as a proved fact." Again," " In all cases, the palea, whatever its origin, is called upon in conjunction with the subtending glume to perform more or less of the functions of the deficient or absent perianth."

It is to be noticed that the male and female reproductive organs occur in the same flower, which is therefore hermaphrodite. The stamens of oats are to be looked upon as metamorphosed leaves, as in the pine ; it seems probable, though not definitely proved, that the pollen sacs are homologous with those of the pine. The immature pollen spore is two-celled, as in most other angiosperms," and it is doubtful whether there is any trace of a prothallium. This stage in the life cycle of the plant, which is so marked in the moss and liverwort, far less prominent in the fern, reduced, if it can be considered present at all, to but a cell or two in the pine, is in this plant probably entirely suppressed.

Attention is called to the fact that the ovule is not

---

[20] Bentham, Notes on **Gramineæ**, Jour. Linn. Soc., xix, p. 23.

[21] L. c., p. 24.

[22] L. c., p. 25.

[23] Cf. Strasburger, **Neue** Untersuchungen, p. 5.

naked as in **pine,** but that **it is** surrounded by an organ
peculiar **to** angiosperms, **the** ovary, which **in this** plant
adheres to the surface **of the** ovule. **It is** much better
developed in Trillium **and** Capsella, to which for **its**
study the student is referred. Since the ovule **is thus**
inclosed, stigmas have been developed as naked pollen-
catching surfaces, to which the pollen spores can
adhere, and through whose loose tissues they can eas-
ily send their tubes. •

The adherence of the ovary wall to the ripened seed
gives **rise to a** fruit peculiar **to** grasses, the grain, which
is commonly mistaken for **a** simple **seed.**

The fibro-vascular system **is** well developed in oats,
and is of the typical monocotyledonous form. The
hypodermal fibers of both **stem and** leaves **give addi-**
tional **strength to these** organs. **The bundles through-**
out the stem are of **the** collateral **type, as in the pine,**
but with this difference ; **whereas in the pine there**
remains a cambium layer between the **xylem and**
phloem, here there is no cambium. The continued
growth of the bundle is therefore impossible, whence
it is known **as a** closed bundle. The axial bundle of the
root is, like that of the **fern, a** radial one. In the **fern**
root a single apical cell **forms** the growing point of
the root ; in oats the apical cell is replaced by a clus-
ter of cells. The remark respecting the root-epidermis
of Adiantum [24] **is equally** applicable here concerning the
older **roots.**

In the leaves the **most novel structure is the groups**
of hygroscopic, or as **they were first named by Duval-**

---

[24] Cf. also Goodale, Physiological Botany, **p. 108.**

Jouve,[15] bulliform cells. They are **found in all** grasses that **roll and** unroll their leaves. **These** cells when they **lose** part of their moisture contract and roll up the leaf, which again expands upon their regaining it. This movement reduces the amount of surface available for evaporation, **and is a** safeguard for the plant. It will be remembered **that the moss** leaf accomplishes **the same result,** but without a specialized apparatus.

---

[15] Étude anatomique de quelques Graminées, 1870, p. 320.

TRILLIUM.

Trillium **recurvatum** Beck.

PRELIMINARY.

THE Trillium designated, as well **as** *T. sessile*, is found in the spring, generally in rich woods, and may be readily recognized **by** the naked stems, from fifteen to thirty centimeters (six inches **to a foot)** or more high, bearing **at the summit a circle of** three broad netted-veined leaves, at the **center of** which (the **apex of the stem)** stands a single sessile dark-purple flower. The **stem rises** from a deep-seated, somewhat toothed, very **thick** rootstock, which bears the fibrous roots **along its** under surface. **In** the other Trilliums, which **are at** all common, **the** flowers **are** usually white **or** pinkish, **or purple in one case,** and stalked. **Any** species may be used for the laboratory work.

Although Trillium **is now** considered a member **of** the lily family, **the** largest order of petaloideous **mon-**ocotyledons, **it is** not a very characteristic member, but has been selected **for its** general distribution, **its** completeness **and simplicity,** and **for** its convenient **size.**

The **materials needed are** fresh **or** alcoholic **spec-**

imens **of roots,** rootstocks, stems, leaves, flowers
and fruit; potassic hydrate; magenta; and iodine.

LABORATORY WORK.

GROSS ANATOMY.

A. GENERAL CHARACTERS. **Note**

1. The *main axis*, **consisting of a** thickened, horizontal,
    under-ground stem, the *rootstock*, **and a** single **vertical**
    branch, the *aerial stem*, bearing a terminal *flower.*

2. The *rootstock*, bearing **as lateral** appendages **the** *roots*,
    and *modified* **leaves** in **the form of broad membranous**
    scales.

3. The *aerial stem*, bearing as lateral appendages **a whorl**
    of three *leaves*, and the parts of the *flower.*

B. THE ROOTS. **N**ote

1. Their *arrangement* on the rootstock.

2. The almost entire **absence of** *branching*.

3. The *surface*, especially the transverse wrinkles **on older**
    parts.

4. The *root-hairs.*

Mount a transverse section of the proximal portion of **a**
large root, and notice

5. The presence and **relative** areas of *three regions:*

    a.   The *cortical region* of two layers :
        i.   An *outer layer* of small **cells.**
        ii. An *inner layer* of large, irregular, loose cells,
             often torn through in sectioning, the unequal

development of which gives rise to **the** wrinkling.

b. **The** median large-celled or *parenchyma region.*

c. The central or *fibro-vascular region,* **in** which may be detected the large openings of four or **five** tracheary vessels.

Mount a longitudinal section through the center of the **root and notice**

6. **The several** regions **as** before.

7. The depth **of** the surface wrinkles.

8. **Diagram both** the transverse and longitudinal sections.

C. THE ROOTSTOCK (Subterranean Stem). Note

1. **The** *shape* **and** *thickness.*

2. The succession of *nodes* **and** *internodes.*

a. The *number* **in an** unbroken rootstock.

b. The *scars* **of** former branches, **and the** varying number of intervening nodes.

c. The *irregular growth* of **the** internodes.

3. **The** prevailing number of roots from each node.

Mount a transverse section ; notice

4. **The** *three* **parts :**

a. **The** extremely **narrow** brownish *cortical region,* **forming the boundary** of the section.

b. **The great** mass **of** the stem, **whitened with the** reserve food material. In thin **parts of the sec-** tion, where the food material has been washed out in mounting, note

i. The delicate colorless tissue, *fundamental par- enchyma.*

Tease a bit of the stem in a drop of water on another slide, treat with iodine, and note

> ii. The color imparted to the **food material,** indicating its nature.

> *c.* Scattered **irregularly** through **the section** the comparatively **few small dark areas, the** *fibro-vascular bundles.*

Make a longitudinal section, and notice

5. **The three parts seen in the** transverse section ; the bundles **branching** irregularly and obscurely.

6. The *growing apex.* Note

> *a.* The *shape.*
> *b.* The **sheathing membranous** *scales.*
> *c.* The position **of the aerial branch.**

Bisect the apex, and upon the cut surface notice

> *d.* The two or more rather thick projecting **bracts.**
> *e.* Beneath these a very small protuberance, the growing point of **the** stem, sometimes **accompanied by** smaller **lateral** protuberances, **the rudimentary** scales.
> *f.* **Draw the bisected apex.**

D. THE BRANCH (Aerial Stem). Note

1. **The absence of nodes** below **the whorl of leaves.**

2. **The** smoothness of the *surface.*

Mount a transverse section. **Notice**

3. The *three parts :*

> *a.* The very narrow *cortical region.*
> *b.* The *fundamental parenchyma,* forming **the** ground work of the section ; the cells appearing empty.
> *c.* The limited number of *fibro-vascular bundles;* note
> > i. Their *arrangement* and relative *size.*

      ii.   The two areas in each : the light colored por-
tion lying toward the outside of the stem, the
phloem, the **shaded** portion **lying toward the**
center of the stem, the *xylem*.

4. Draw the section.

5. Make a longitudinal section and note the several parts
seen in the transverse section. Draw.

E. THE LEAF. Note

1. **The** *scale leaves* of the rootstock.

    a.   The bases of decayed scales at each node.

    b.   Younger ones, sheathing the apex of the rootstock
and base of the aerial branch.

Mount a portion of a scale and notice

    c.   **The** parallel veins, **and** intervening parenchyma.

    d.   Numerous dark spots, clusters of **raphides.**

2. The *foliage leaves* of the aerial **stem.**

    a.   The *number* and *arrangement.*

    b.   The *shape.*

    c.   The particular outline of the *apex*, and of the
base.

    d.   The short stalk, **petiole**, if present.

    e.   The distribution of the *veins.*

Mount a transverse section from near the base of the **leaf,**
and note

    f.   **The several** parts :

       i.   The colorless *epidermis.*

       ii.   The *veins*, the largest projecting on the lower
side ; each containing

       iii.   A *fibro-vascular bundle*, having the phloem
area toward the under side of the leaf ; accom-
panied in the largest veins by

    iv. *Colorless parenchyma.*

    v. The darkened (in fresh specimens green) *mes-
ophyll.*

Mount a portion of the epidermis stripped from the upper
surface of the leaf, and beside it a portion from the under
surface. Note

    g. The numerous dots in that from the lower surface,
the *stomata.*

    h. The absence of stomata in that from the upper
surface.

F. THE FLOWER. Note

1. The several parts arranged in whorls.

    a. The outer whorl of three **sepals,** constituting the
calyx.

    b. The second whorl of three **petals,** constituting the
corolla.

    c. The third and fourth whorls of three stamens each,
constituting the **androecium.**

    d. The innermost whorl of three partly united carpels,
constituting the **gynoecium.**

2. That the parts of the flower arise from the broadened
extremity of the stem, the *receptacle.*

3. The *alternation of the parts* of each whorl with those of
the whorl next to it.

4. The *sepals.*

    a. The *shape.*

    b. The *color* in a fresh specimen.

    c. The *venation.*

    d. Draw a single sepal.

5. The *petals.*

   a.   The *shape.*

   b.   The *color* in a fresh specimen.

   c.   The *venation.*

   d.   Draw a single **petal.**

6. The *stamens.*

   a.   The several parts of each.

       i.   The stalk or *filament,* passing into

      ii.   The *connective,* on the right and left **margins of** which are borne

     iii.   The pollen sacs, or *thecæ ;* the connective and thecæ together constituting the *anther.*

   b.   Draw a stamen.

Mount a transverse section of the filament, along with one of the anther, and **note**

   c   The *filament.*

       i.   **The** *outline.*

      ii.   The uniform *ground tissue,* **containing a cen-**tral *fibro-vascular bundle.*

     iii.   Draw.

   d.   The *anther.*

       i.   The outline of the *connective,* and character of its tissues.

      ii.   The form of the *thecæ.* The mode of bursting is often well shown.

     iii.   Draw the section

   e.   The *pollen,* escaped from the thecæ and appearing **as** fine particles. Dust some from **an anther and examine dry, noting color** and pulverulence.

7. The *carpels.*

   a.   The **parts of each.**

       i.   **The enlarged basal** portion, bearing a pair of **prominent ridges,** united with the bases of the **other carpels to** form the *compound ovary.*

ii.   The tapering divergent *styles.*
iii. The double wavy crest along the inner face of each style, the *stigma.*
b.   Draw the three carpels.

Make a transverse section through the middle of the *style ;* mount, and notice

c.   The *outline,* including the stigma.
d.   The central *fibro-vascular bundle.*
e.   The *stigma.*
    i.   The recurved *sides,* and deep median groove extending in to the central bundle.
    ii. The velvety *papillæ* clothing its surface.
f.   Draw the section.

Mount several transverse sections of varying thickness passing through the middle of the compound ovary, and notice

g.   The three similar parts belonging to the three carpels of which it is composed, each consisting of
    i.   The two prominent ridges or *wings.*
    ii. The *fibro-vascular bundle* lying in the tissue between the ridges.
    iii. The sides of the carpel uniting with the sides of the adjoining carpels, extending into the cavity of the ovary and meeting in the center, forming the three placentæ.
    iv. The two or more *fibro-vascular bundles of the placentæ.*
    v.   The rounded *ovules* in the cavity of each carpel, borne on the sides of the right and left placentæ.
h.   Draw the section.

G. THE FRUIT. Notice

1. The *sepals* still in growing condition.
2. The withered but persistent *petals, stamens* **and** *styles.*
3. The fleshy winged **pod, like the** young **ovary, but** larger, inclosing
4. The *seeds.* Note
   a. The slender *attachment.*
   b. The fleshy body on the side of the seed, the **strophiole,** which is an outgrowth of the lower part of
   c. The stalk of the seed, **funiculus,** extending beyond the strophiole as a ridge on the seed, the **rhaphe,** and terminating at
   d. The base of the seed, the **chalaza.**
   e. The *shape.*
   f. The minutely granular *surface.*
   g. **Draw a seed.**

Beginning at the **chalaza, cut several** very thin **transverse** sections, then **another transverse section from the** middle of the seed, mount, **and notice**

   h. The thin brown coat of the seed, *testa.*
   i. The uniform tissue within, the cells of which are filled with reserve food material, and in the section from the middle of the seed are seen to radiate from the center to the outside.
   j. In the sections from the base of the seed, the small round spot, between the center of **the** section and the **side** next the rhaphe, the *embryo.*

MINUTE **ANATOMY.**

A. THE **ROOT. In a central** longitudinal **section** through the **root-tip, note** under low power

1. The outer looser cells and inner more compact tissue forming the **root-cap.**[1]

---

[1] It may be difficult **to** get the region complete.

2. A cluster of small angular cells in the center of the section just behind the root-cap, the *primary meristem*, forming the growing point.

3. Originating in the primary meristem, an epidermis-like row of cells of the root proper, the *dermatogen*, which is continuous with the epidermis of the surface of the root.

4. The central cylindrical mass of cells, to become the fibro-vascular column, the *plerome*.

5. The region of cells between the plerome and dermatogen or epidermis, to become the cortical portion of the root, the *periblem*.

6. The branching *root-hairs* and their relation to the epidermal cells.

7. Draw the section, showing the regions above noted.

Make a transverse section of the root at its wrinkled part. Under low power, notice

8. The peripheral portion of two or more rows of rounded cells.

9. The underlying large-celled tissue, with delicate distorted walls, often torn in sectioning.

10. The outer portion of the core, with rounded cells becoming smaller as they lie nearer

11. The circular axial part of the core, the fibro-vascular bundle, containing several large vessels arranged in four or more short radiating rows.

Use a high power, and notice

12. The surface row of cells, the *epidermis*, slightly or not at all differing from

13. The cells beneath the epidermis, the *hypoderma*.

14. The very thin walls of the *loose tissue*.

15. The thickened walls of the outer part of the **core**, and the intercellular spaces at the angles.

16. Draw a portion of these several tissues of the root.

17. The somewhat irregular **row** of thickened, slightly colored cells, the *bundle sheath*, usually with an evident *middle lamella* in the walls.

18. The *fibro-vascular bundle*.

    a. Immediately within the bundle-sheath a row of thin-walled cells, the *pericambium*.[2]

    b. The radial rows of large *vessels*, larger toward the center, smaller toward the periphery.

    c. The intermediate groups of *sieve tissue*.

    d. The thin-walled cells of the center of the bundle.

    e. Draw the fibro-vascular bundle and bundle-sheath.

Make several longitudinal sections through the core of the root ; identify and study

19. The several tissues seen in transverse section, **drawing** a few cells of each.

B. THE ROOTSTOCK (Subterranean Stem). Make a transverse section, mount in strong potassic hydrate, and notice

1. **The row** of *epidermal cells* with brown outer walls.

2. **The** *parenchyma* **tissue** within, filled with starch, **and forming the mass of the** rootstock.

3. **Draw a few** cells of **epidermis, and of adjoining paren-chyma.**

4. The *fibro-vascular bundles*, cut at all angles, even longi-

---

[2] Sometimes absent in old roots by having become permanent tissue.

tudinally. Owing to the difficulty of obtaining good sections, and the interference of the starch, the further study of the bundles is deferred till they are reached in the aerial stem.

Cut a thin slice from the surface of the rootstock, and under low power, notice

5. The shape of the epidermal cells, and absence of sto-mata. Draw.

Make a longitudinal section through the growing tip in the plane of the branch, and note

6. The *sheath* composed of one or more sets of thickened and partly coalesced bracts. These bracts may be detected in various stages of growth, all originating behind

7. The *growing point.*

8. The growing tips of *rudimentary roots and branches.* Note the exogenous development of a branch, all the tissues of the stem entering into it, and the endogenous devel-opment of a root, distinct from the tissues of the stem and pushing its way through them.

C. THE BRANCH (Aerial Stem). In a transverse section, with a low power, notice

1. The single row of *epidermal cells.*

2. The loose, round-celled *fundamental parenchyma,* with large intercellular spaces.

3. The *fibro-vascular bundles,* consisting of

    a.   The light colored portion, *phloem.*
    b.   The shaded portion, *xylem.*

   Under high power notice

4. The *epidermis* and *hypoderma.*   Note

a. **The** thickness of the different **walls.**

b. Draw.

5. The *parenchyma* of the center **of** the **stem.** Note

a. The shape of the cells.

b. The thinness of the walls.

c. The large *intercellular spaces.*

d. Draw.

6. The *fibro-vascular bundle.* Note

a. **The** *phloem,* consisting of angular cells of various sizes; the smaller, *conducting* **cells** or *bast parenchyma*, the larger, the *sieve tissue.*

b. The *xylem,* in the smaller bundles of less area than the phloem, but in larger bundles greater, and spreading out on either side the phloem, until, **in some cases, it entirely** surrounds it.[2] **It is made** up of

  i. The *vessels* of various sizes, and

  ii. Interspersed among the vessels and extending out on the axial **side,** rather angular cells, the *wood parenchyma.*

7. The absence of meristem cells between **xylem and** phloem, thus forming **a** *closed bundle.*

8. **Draw a** small and a large bundle.

Take a longitudinal section through the center **of the stem** and notice

9. The several **tissues seen in transverse section.**

a. The epidermal and hypodermal **tissues. Draw.**

b. The central parenchyma. Draw.

---

[2] Forming the external sheath (Prantl and Vines, Text-book, p. 58), which is probably but a special development of the surrounding cells of the fundamental system.

c. The several tissues of the *bundle.*

 i. The *phloem,* of long, thin-walled, uniform cells. Draw.

 ii. The *vessels,* of various sizes and kinds. Draw some of each.

 iii. The *wood parenchyma,* considerably like the bast parenchyma, but in its best development with thicker walls, and more evident somewhat oblique end walls. Draw.

Strip some epidermis from the surface of the stem, and under low power, notice

10. The shape and arrangement of the cells, and of the stomata. Draw.

D. THE LEAF. Make a vertical section near the middle of a foliage leaf, at right angles to the principal veins, and notice

1. The row of colorless *epidermal cells* bounding the upper and lower surfaces.

2. Here and there transverse sections of the *veins.* Note the depression above, and the convexity below the larger veins.

3. The *mesophyll.* Note

 a. The layer of closely set cells along the upper side, *palisade parenchyma.*

 b. The loose irregular tissue on the lower side, *spongy parenchyma.*

Under high power, notice

 c. The shape and position of the cells of the palisade parenchyma.

 d. The shape of the cells of the spongy parenchyma and the numerous intercellular spaces.

 e. Contents of the mesophyll cells.

4. The *epidermis.* Note

   a. The shape and irregularity of the cells on the *upper side* of the leaf.

   b. The shape and irregularity of the cells on the *lower side*, together with the shape of

   c. The numerous *stomata* in transverse section, when cut through the middle, and when through the ends.

5. Draw a part of the mesophyll and epidermis, extending from one side of the leaf to the other.

6. The *veins.* Selecting a vein of moderate size, note

   a. The small *epidermal cells* on the upper side, and the large ones on the lower side, both more thickened and regular than elsewhere on the leaf.

   b. The *fibro-vascular bundle*, with the xylem area toward the upper side, and the phloem area toward the lower side[4] ; sometimes with

   c. *Fundamental parenchyma* interposed between the bundle and the epidermis on either side.

   d. Draw a vein.

Strip some epidermis from the upper side of the leaf, and some from the lower side, and notice

7. The shape and arrangement of the cells of the two surfaces, and the abundance of stomata. Draw.

E. THE FLOWER.

1. The *sepals.* Make a transverse section through the middle of a sepal, and notice

   a. The several parts and tissues seen in the leaf. Draw.

[4] Note how this follows from the relative positions of xylem and phloem in the stem.

Strip some *epidermis* from the upper and lower sides, and notice

  b. The shape of the cells, and presence or absence of stomata in each. **Draw.**

2. The *petals.* **Make a** transverse section **through the** middle of a petal, **and** notice

  a. The several parts seen in the leaf. **Draw.**

Strip some epidermis from **the** upper and lower sides, **and notice**

  b. The shape of the cells and presence or absence of stomata in each. **Draw.**

3. The *stamens.* Make a **transverse** section through the middle of a filament, and at the same time through the middle of the anther, and notice

  a. The parts of the *filament.*
    i. The *epidermis*, with a minutely **sinuous** outline to the free surface.
    ii. The *parenchyma* beneath.
    iii. The central *fibro-vascular bundle*, consisting **largely** of xylem, with only a few small groups **of scattered phloem** cells **on** the more convex (**outer**) side.

  b. **Draw part of** the filament.

  c. **The** *connective*, with tissues **like those of the** filament.

  d. The *theca*, consisting **of**
    i. The two *valves* usually broken away **at the** tips, springing from
    ii. The *base* which merges into the connective.

  e. The *wall* of the thecal valves. **Note**
    i. The *epidermis*, like that **of the** filament but **with** numerous *stomata*, seen in transverse section.

·ii. The *endothecium*, having its cells provided with transverse thickenings.

iii. The *broken cells* at the tips, corresponding to the broken cells at the middle of the **base**.

f. The base of the theca. Note

i. The arrangement of the spiral cells.

ii. The median cavity and its extent.

g. Draw a theca in outline, and fill in a portion of the tissues.

h. The *pollen spores.* Note

i. The *surface* of the wall, its roughness, **and** the simple, large, round or oblong spots to be seen upon some spores.

ii. The wall in *optical section.*

iii. Draw.

Burst some spores by pressing on the **cover glass. Note**

iv. The empty outer layer of the wall, *extine.*

v. The remainder **of** the spore, still enveloped by the thin *intine,* roughened like the extine.

vi. The *contents* from broken spores, with minute granules.

4. **The** *carpels.* In a transverse section through the style, **notice**

a. The *epidermis* and *parenchyma* like that of the **fila-ment** of the stamen.

b. **The** *fibro-vascular bundle* with apparently **no** phloem, elongated to embrace the cleft of the stigma.

c. The *stigmatic surfaces,* covered with papillæ.

d. **In** some of **the pollen** spores lodged upon the papillæ, **the** developing pollen tubes may be detected.

e. Illustrate with drawings.

In a transverse section through the *compound ovary* notice, under low power,

- *f.* 　The uniform, continuous *epidermis.*
- *g.* 　The *loose parenchyma* tissue of the wings and walls.
- *h.* 　The *fibro-vascular bundles* under the sinus between each pair of wings, and the groups in the placentæ.　The latter give off branches to
- *i.* 　The *ovules*, placed back to back in the ovarian cavities.　Note
  - i.　The thick stalk, *funiculus*, joined to the side of the ovule, quite inverting it (*anatropous*).
  - ii.　The *two coats of the ovule*, the inner protruding from the outer.
  - iii.　Draw.
  - iv.　If a section passing through the center of an ovule has been obtained, note the *nucellus* within the inner coat, and
  - v.　In the center of the nucellus the small *embryo sac.*
  - vi.　Draw.

Under high power, notice

- *j.* 　The epidermis and loose parenchyma at the base of a wing.　Draw.
- *k.* 　The fibro-vascular bundle of the wall, showing a little phloem only on the outer side.　Draw.
- *l.* 　The parenchyma and fibro-vascular bundles of the placentæ.　Draw a part.
- *m.* 　Where the section has passed through the center of an ovule, note the similarity of the cells of all the parts.　Draw.

F. THE FRUIT.　So little change has taken place in the growth of the fruit that it is only necessary to study the minute anatomy of

1. The *seed.* **Make** several sections in succession from the base of the **seed, and** an additional section from the middle, all at right angles to the longer axis **of the** seed. Note in the latter, under low power,

    a. The *colorless tissue,* filled with reserve *food mate-rial,* radiating from the center of the seed.
    b. **The** thin brown *testa.*
    c. **The** large-celled tissue of the *strophiole,* through the middle of which passes a fibro-vascular bundle belonging to the funiculus.

In the sections from the base of the seed, note

    d. The group of small cells lying in a cavity **on one** side the center, the section of the *embryo.*

Under high power, notice

    e. The several *sections of the embryo.* **Draw.**

In the section from the middle of the seed, **notice**

    f. The colorless tissues **of the center.**
    g. The tissues **of** the testa.
        i. The outer layer of oblong cells.
        ii. An inner layer of more elongated cells.
        iii. Within this, very delicate cells, not easily made out.
    h. Draw some of the colorless cells, with adjoining testa.
    i. The tissue **of the** strophiole. Draw.

Take a surface slice from the seed, and notice

    j. The shape **of** the *surface cells.* **Draw.**

ANNOTATIONS.

The most notable **advance** in Trillium, so far as its gross anatomy is concerned, is the development of a complete flower.[5] The so-called flower óf Atrichum differs essentially from a true flower in bearing the primary sexual organs (antheridia and archegonia) directly upon the axis of the flower. In true flowers asexual spore structures first arise in the shape of pollen spores and embryo sac, which in turn give rise to the primitive sexual organs as naked cells within these. The first true flowers are met with in Pinus, in which they have no envelopes; in Avena there is either no perianth, or a very poorly developed one, just as we choose to regard the palet and lodicules as such or not; but in Trillium there is a typical perianth. The flower is composed of sets of modified leaves symmetrically clustered at the apex of a short axis. The outermost and lowest whorl (calyx) of leaf-like sepals, a second whorl (corolla) of colored petals, then two whorls of stamens (andrœcium), and a central and uppermost whorl of carpels (gynœcium), all standing upon the broadened apex of a branch (receptacle) may be taken to fairly represent a typical flower. The order given not only expresses the order of occurrence upon the receptacle but also the order of development; and the type number, three, so characteristic of monocotyledens, should be borne in mind (see fig. 1).

The perianth is an arrangement for protection both in bud and blossom,[6] while in the latter stage the corolla becomes in addition a device for attraction.[7]

---

[5] For a concise account of the homology and nomenclature of the parts of a flower see Sachs, Text-book, 2nd Eng. ed., p. 490.

[6] Cf. Kerner, Flowers and their Unbidden Guests.

[7] Cf. Gray, Struct. Bot., p. 215, et seq.: Darwin, Fert. of Orchids; Effects of Cross and Self-fertilization; Forms of Flowers etc.

While this is generally true throughout phanerogams, there **are cases** in **which the** calyx becomes attractive, **and** other cases in which the calyx and corolla have **so far** lost their original functions that they neither attract nor protect.

In the blended carpels, forming the pistil which in-closes the ovules, we have the characteristic feature of angiosperms as distinguished from gymnosperms. In **Avena** the ovule is adherent to the wall of the ovary, but in Trillium it is distinct; the whole after fertilization forming the fruit with its inclosed seeds. Each carpel is an infolded leaf, bearing the ovules upon its edges (see fig. 11). Each should normally, then, contain **two** rows of ovules, corresponding to the right and left margins of **the** leaf, as in fact is the case in Trillium. **These lines of** attachment, **the theoretical leaf** margins, are known as placentæ. The upper **part of the carpel-**lary leaf is generally modified **to form a long or** short style, while the stigma is a surface formed of cells se-creting a viscid fluid, and more or less modified for **the** reception and retention of pollen.[8]

The nucellus is the part of the ovule to appear first, **followed by** the inner and outer integuments in the order named (basipetal). Concerning the homology of the ovule **there has been and** still is much discussion, and **the student desiring to pursue the** subject must look **to its** extensive **literature.**[9] **The** anatropous **ovule of Tril-**

---

[8] For discussion of the pistil **and** carpel, and references to the litera-ture of the subject, see Gray, Struct. Bot., pp. 166 (with footnotes), **259,** et seq.

[9] Gray, Struct. Bot., pp. **267,** 282 ; Eichler, Blüthendiagramme, **part** II, page xv; Warming, **De** l'Ovule, Ann. Sci. Nat., sér. 6, v, p. **177 ;** Van Tieghem, op. cit., **sér. 5,** xii, p. **312 ,** Sachs, Text-book, 2nd Eng. ed., pp. 492, 570, etc.

lium may be mentioned as by far the most common form.

As in Adiantum, the main axis of Trillium is subterranean, horizontal and thickened, forming a rootstock, a protective measure which in this case is correlated with spring blooming and a long period of rest. The aërial part of Adiantum, however, is a leaf, while in Trillium it is a branch, the " root-leaves " being reduced to membranous scales on the rootstock. A single branch (rarely two) is sent above ground each season, and a series of corresponding annual scars may frequently be seen upon the stem. In the meantime the terminal bud of the latter continues to develop and to thrust its way through the soil, protected in this root-like habit by a special modification of bud-scales.

Probably the most exceptional character of Trillium is the venation of its foliage leaves, which is of the netted type, instead of the parallel venation most characteristic of monocotyledons, as in Avena. Among net-veined leaves they are palmate, a type which produces a broad expanse of surface, very favorable for the accomplishment of leaf work. In this way Trillium has secured a large exposure of surface to air and sunlight for a plant so low in stature, and generally deeply sunk in vegetable debris.

The primary root-structure is quite uniform in all plants, consisting of epidermis (or a piliferous layer), cortical parenchyma, and a central fibro-vascular cylinder. In the central cylinder xylem and phloem masses alternate with each other, the intervening spaces being occupied by parenchyma (forming a radial bundle); surrounding all is a layer of parenchyma, the pericambium, and outside of this a single layer of modified cortical cells, the bundle sheath or endodermis. Rootlets of most cryptogams originate from the bundle sheath, while in phanerogams they usually come from the pericambium.[10] In the radiating lines of tracheary vessels it will be noticed that the larger vessels are toward the center, the smaller spiral vessels being peripherally placed, thus reversing the order of the stem.[11]

The stem is very much modified by becoming a food reservoir. The epidermis is poorly limited, often scarcely distinguishable from the adjacent cortical parenchyma ; the parenchyma of the fundamental system is greatly developed and filled with starch ; while the fibro-vascular bundles are reduced and irregular in their course. In most cases the bundles bend outwards, and such are evidently leaf-traces,[12] though they have lost all definite connection with the accompanying scale-like leaves.

The terminal bud of the stem is large and its parts very distinct. It is taken to represent in our series a typical terminal bud of phanerogams. A section reveals the fact that the growing point (*punctum vegetationis*) consists of a group of cells, in place of the single apical cell of cryptogams, and the forming tissues diverge from it in well defined lines. The three typical regions of the stem originate from different regions of this primary meristem, so that they do not

---

[10] Prantl and Vines, Text-book, p. 51; Strasburger, Bot. Pract., p. 276.

[11] Cf. Goodale, Physiol. Bot., p. 110, et seq.; Prantl and Vines, Text-book, p. 51, DeBary, Comp. Anat., p. 348.

[12] Cf. Prantl and Vines, Text-book, p. 46; Goodale, Physiol. Bot., p. 125; DeBary, Comp. Anat., p. 233.

have a common origin, but are distinct from the first. Lateral members soon appear as small protuberances, which are rudimentary leaves, branches, or roots, those nearest the apex being the youngest. The **axis** remains so **short that** the **lower** leaves **overlap** the growing point, and in the case of this underground stem thicken and coalesce at their tips, forming a continuous and firm sheath, thus performing the same office of pro-**tection for** the rootstock that the root-cap does for roots. In the same Trillium bud may be seen the essentially different mode of development of the root and branch, the former being endogenous and pushing its way through the overlying tissues, the latter exog-enous,[13] blending with the surface tissues of the stem.

The branch, or aerial stem, is remarkable for its single long internode. The fibro-vascular bundles are rather poorly developed, the monocotyledonous stem bundle being more typically represented in Avena. In Trillium, however, **are** seen such monocotyledonous characters **as** the isolation **of the** bundles within an **abundant** fundamental **parenchyma**, the well-marked phloem and xylem areas with no intervening cambium (closed bundles), and an occasional development of an external sheath.[14]

The histology **of the** leaf **is in** general much the **same as** in the majority **of** leaves ; it differs from Avena in the presence of a much better developed mesophyll. A broader expanse of **leaf tissue** necessitates a greater branching of the supporting and conducting fibro-vas-cular system, and **a better** differentiation of leaf sur-

---

[13] Prantl and Vines, op. cit., p. 23.
[14] Prantl and Vines, op. cit., p. 58.

faces. **Hence** the palisade and spongy parenchyma are quite distinct, and the large intercellular spaces of **the latter** have more frequent connection **with the** outer air through stomata. The fact that in the fibro-vascular bundles of leaves the xylem is always on the upper side, and the phloem on the lower, should **be** recognized **as** necessitated by the relation which **they hold to** the same regions in the stem.

In the ovary the distribution of **fibro-vascular bundles and the** relative positions of phloem and **xylem** are most interesting and suggestive. **The** single **bundles** in the outer walls have the phloem toward the outside (the lower surface of the carpellary leaf), and the xylem toward the inside. But the inner group of bundles in the ovarian partitions have their arrangement reversed, the phloem being **on the** inside (toward the center of the flower) and **the** xylem **on** the outside [16] (see fig. 10). This is readily accounted **for by the** infolding of the carpellary leaves (see fig. 11).

The walls of the ovary show also a region of very **loose** spongy tissue, developed in the mesophyll of the carpellary leaf, and extending into the style, the conducting tissue, " which serves as a path of least resistance **for** the penetrating **pollen tubes.**" [16]

---

[16] Cf. Goodale, Physiological Botany, p. 173.

[16] Goodale, Physiol. Bot., p. 172.

SHEPHERD'S PURSE.

Capsella Bursa-pastoris Mœnch.

PRELIMINARY.

THIS plant, chosen to represent the highest development of plant life, is of European origin, but has become abundant in this country and elsewhere, being one of those vigorous foreign species which hasten to take possession of any cleared or cultivated land. It is found everywhere around dwellings, and in fields and waste grounds. It has not only the advantage of universal distribution, but also of continuous growth throughout most of the year, even a few warm days in winter calling it into bloom. One who does not already know the plant can recognize it from the fact that it is a low, quite insignificant herb, varying in height from five to fifty centimeters (two inches to a foot or two), with a rosette of rather narrow jagged root-leaves often lying flat upon the ground, much smaller scattered stem-leaves, and very small white flowers constantly opening at the summit of the branches, and producing small triangular rather heart-shaped pods below (see fig. 6). It is most abundant in spring and early summer, but can be found in bloom throughout the warm months, and may be grown in the green-house for use in winter.

The materials needed for studying are alcoholic (or fresh) specimens of roots, stems, leaves, flowers, and pods ; fresh specimens of **flower buds ; and** potassic hydrate.

LABORATORY WORK.

GROSS ANATOMY.

A. GENERAL CHARACTERS. Note

1. The main axis, consisting **of a** *root* and *stem*.

2. Three kinds of *branches :*

   a. One on the root, *rootlets*.

   b. Another on the *stem* and simply repeating it, long or very short, or represented by small buds.

   c. **The third on** the upper part of the stem, **each** bearing a single **flower, pedicels.**

3. The position of **the stem branches with reference to** the stem leaves, *axillary*.

4. The *nodes* and *internodes* **of** the stem, as indicated by **the** insertion of the leaves.

5. The absence of *lateral appendages* on the root or its branches ; those of the stem and its branches appearing as foliage and flower parts.

6. **The absence of** leaves or bracts subtending the pedicels.[1]

B. **THE ROOT. Clean** thoroughly, **immerse in water** over a **dark surface, and note**

1. The arrangement **of the** *branches* (*rhizotaxy*).

2. The thickened whitish **tips of uninjured rootlets.**

---

[1] A notable peculiarity **of** the order *Cruciferæ*, of which Capsella is a member.

3. *Color* as contrasted with that of the growing stem.

4. The *root-hairs* near the tips of rootlets.

Make transverse and longitudinal sections of a medium sized root and note the presence and relative importance of

5. The *three tissue regions :*

    a. The thin peripheral or *cortical region.*

    b. The large axial or *central cylinder,* in which radiating lines formed by large ducts can usually be seen in the transverse sections.

    c. A region of loose *colorless cells* between the other two regions.

 Peel the outer layers from a branching root, and notice

6. The axis of each rootlet remains attached to the axis of the main root.

C. THE STEM. Note

   1. Mode of *branching.*

   2. Surface *markings.*

   3. The relative lengths of *internodes.*

   4. Axillary *branches* or *buds.*

Make a transverse section and note

5. *Three regions :*

    a. The peripheral or *cortical region.*

    b. The narrow median or *fibro-vascular region.*

    c. The axial or *pith region.*[2]

6. The *fibro-vascular bundles.* Note

    a. *Shape* and relative *size.*

    b. The cut ends of the tracheary *vessels,* as holes through the bundles.

---

[2] Not present in the root.

7. Draw the section.

Make a longitudinal section through **a branch and** leaf-bearing node, and note

8. The three regions, as well as

9. Their relation to the leaf and branch.

10. Illustrate with diagram.

D. THE LEAF. Note

1. Two sorts of leaves :

   a. *Root-leaves,* clustered at the base of the stem.
   b. *Stem-leaves.*

2. Leaf *arrangement* (*phyllotaxy*). Observe that an imaginary line connecting the insertions of successive stem leaves is a spiral. Discover the number of times the spiral encircles the stem, and the number of leaves it passes, before reaching a leaf standing directly over the first.[3]

3. Leaf parts ; in the root-leaves a blade and leaf-stalk or petiole, in the stem-leaves simply a sessile blade.

4. Leaf *shapes* and *sizes,* the great variety. Draw a series of the most characteristic.

5. Leaf *surfaces ;* differences between the upper and lower. Notice
   a. Simple hairs.
   b. Stellate hairs.

6. Distribution of the *veins,* and their relation to the teeth.

7. The uncoiling of the *spiral threads,* when the leaves are broken by careful stretching.

---

[3] The student may find it easier to substitute a thread for the imaginary line, and must also allow for any twisting of the stem.

E. THE FLOWER. Note

1. The *four sets of organs* and the **number of** parts in each.

2. The *receptacle*, the enlarged end of the stem.

3. The *sepals*.
   - a.   The *number* of whorls.
   - b.   The *shape*.
   - c.   The *color* in fresh specimens.
   - d.   Draw a single sepal.

4. The *petals*.
   - a.   The *number* of whorls.
   - b.   The *shape*.
   - c.   The *color* in fresh specimens.
   - d.   Draw a single petal.

5. The *stamens*.
   - a.   The *number* of whorls, and number of stamens in each.
   - b.   The *lengths* and *positions* compared with one another.
   - c.   The position of the long pairs and short single stamens with reference to the petals.
   - d.   The four greenish elevations on the receptacle, **nectaries,** alternating with the paired and single stamens.
   - e.   The *filament,* to the tip of which is attached
   - f.   The *anther.* Note
     - i.   The two *thecæ.*
     - ii.   The very narrow *connective.*
     - iii.   The lines of *dehiscence* (best seen in the anther of an advanced bud).
     - iv.   The shape of *apex* and *base.*
     - v.   The different appearance of the inner and outer faces.

vi. With a needle gently break open the thecæ, and, mounting dry, note the abundance, color, and pulverulence of the *pollen.*

g. Draw an uninjured stamen.

6. The *pistil.* **In** a just opened flower, note

a. **The** *shape.*
b. **The three** *parts,* stigma, style, and **ovary.**
c. **The** *ovules.*
d. **Draw an** uninjured pistil.

7. Construct a diagram (like fig. 1) to show the relation **of** the parts of the flower to each other.

F. **THE** FRUIT. Taking the oldest well-formed pod, note

1. **The** *shape.*

2. The *median ridge* of each flattened **face extending into** the persistent style.

In a transverse section **of a pod note**

3. The *partition* formed by the inward projection of the **two** placentæ.

Open a pod by pulling away the two *valves* from the ridge, and note

4. **The** *shape* **of the valves.**

5. The membranous *partition.*

6. **The** *seeds.* Note

a. **The** *funiculus* and its attachment.
b. The roughened *surface,* best seen **in dry** seeds.
c. The *shape,* appearing curved **upon itself,** *campylotropous.*
d. Draw **a** seed showing its attachment.

At the free end of as old a seed as possible, tear the seed-coats slightly, and by careful pressure force out

e. The *embryo*. Make out the relations of

    i. The seed-leaves, the *cotyledons*.

    ii. The initial stem, the *caulicle*.

f. Draw the embryo.

In a transverse section across the long axis of the seed, note

g. The *seed-coats*.

h. The cut faces of the two cotyledons, rather flat and in contact with each other.

i. A section of the caulicle, at the narrower end of the section, and lying against the back of one cotyledon (*incumbent*).

j. Diagram the section.

MINUTE ANATOMY.

A. THE ROOT.

1. The *rootlets*. Remove some of the smallest rootlets, examine with a high power, and note

    a. In complete rootlets, the *root-cap*.

    b. Behind the root-cap the central cluster of *meristematic cells*, from which diverging rows of cells pass off, gradually merging into

    c. The *permanent tissues* of the rootlet.

    d. The *root-hairs ;* the portion of the root on which found.

    e. Make a diagram of the tissues of the root-tip, and also draw some root-hairs accurately.

Make a transverse section of a very small rootlet,[4] examine

---

[4] Secondary changes take place so rapidly in the roots of Capsella, that some difficulty may be experienced in finding one with typical primary root structure ; very young rootlets should therefore be selected for examination.

with a high power, and note

f. A *peripheral region*, consisting of the epidermis,[5] and a few layers of hypodermal cells.

g. A *median region* of colorless cells, the **cortex**.

h. Within this a single row composed of somewhat smaller, **more** closely packed cells, the *bundle sheath*.

i. The space inclosed by the sheath entirely occupied by the *fibro-vascular region*, in which note

    i. The two *xylem masses*, placed with their inner ends of **larger cells in** contact, **thus** dividing the region into similar halves (*binary*).

    ii. The two *phloem masses*, placed right and left of the xylem.

    iii. A limited amount of *parenchyma*, separating the xylem and phloem portions from each other, and both from the bundle sheath.

j. Draw **the** whole **section.**

A series of transverse sections of **larger and larger** rootlets will show *secondary changes*, characteristic **of dicoty**ledons, up **to**

2. The *main shaft*. Make **a transverse** section, examine with a low power, and note

a. The entire disappearance of epidermis.

b. The *cortical parenchyma*.

c. The *fibro-vascular cylinder*.

Using a high power, note

d. The *cortical parenchyma* arranged in radiating lines, **from the** tangential and radial division **of its** meristem cells. The outer **layers** may be composed of thin-walled, close **set, squarish** cells, **cork**; within which are often other **layers of larger cells,** much distorted from the pressure under which the

---

[5] The epidermis may have disappeared even in a very small rootlet.

root grows. Any of the cortical cells may have thickened walls.

e. The alternating *xylem* and *phloem masses.* In comparison with the sections under A. 1. note

    i. The increased number and irregularity of xylem and phloem rows, making longer and shorter radial rows, quite obscuring the primitive arrangement.

    ii. An almost complete disappearance of intervening parenchyma, xylem and phloem being in close contact.

f. Draw a segment showing the tissues noted.

Make a radial longitudinal section, and using a high power,

g. Compare thoroughly with the regions of the transverse section.[*]

h. Illustrate with drawings.

B. THE STEM. Make a transverse section, examine with low power, and note

1. Three regions :

    a. Cortical region.
    b. Fibro-vascular region.
    c. Pith.

2. The *fibro-vascular bundles ;* shape and relative sizes.

Examine with a high power, and note

3. One row of *epidermal cells,* with very thick outer and inner walls. Note

    a. The cuticle.

---

[*] Several sections will be needed to show all the regions of the fibro-vascular area.

    b.   Sections of stomata.

4. **Several rows of** *parenchyma cells.*

5. Usually one row of tolerably large, nearly empty parenchyma cells, the *bundle sheath* (best distinguished **in**
section from growing plant).

6. The masses of *fibrous tissue* rising between the outer
portions **of**

7. The *fibro-vascular bundles ;* selecting one **of the** largest
of these, note

    a.   **The arch** of *bast fibers ;* compare **this** region **with**
the same in smaller bundles.

    b.   Beneath the arch a cluster of smaller irregular cells,
soft bast.

    c.   The *tracheary tissue* and *wood fibers* composing **the**
xylem.

8. The position and extent of the *pith*, **the regularity of**
the cells and the intercellular spaces.

9. **Draw** a segment **of** the section containing **at least one**
bundle.

   Make a radial longitudinal section, including a fibro-vascular bundle,[1] examine with **a** high power and **note**

10. The epidermal cells.

11. **Parenchyma.**

12. **Bundle sheath of** squarish cells.

13. **The bast** fibers, length and nature of **walls.**

14. **Soft bast,** shape **and** walls.

15. Tracheary vessels, variously marked.

---

[1] In a section which passes between fibro-vascular bundles, the same
regions will be noted, except that 13, 14, 15, 16 are replaced by abundant
fibrous tissue.

16. Wood fibers, length and nature of walls.

17. Pith parenchyma, character of walls.

18. **Draw** the various tissues.

C. **THE LEAF.** Make a transverse section of a radical leaf at right angles to the midrib, examine with a high power, and note

1. The *epidermis*, consisting of one row of empty cells on the upper and under surfaces of the leaf. Observe

   a. Sections of *stomata*. **Draw.**
   b. *Hairs* of various kinds. **Draw.**

2. The *mesophyll*, consisting of

   a. An upper row of oblong cells with longer axis at right angles to the plane of the leaf, *palisade parenchyma.*
   b. A lower region of rounded loosely aggregated cells, *spongy parenchyma*, with large and small intercellular spaces, those in connection with the stomata being especially large.

3. The *fibro-vascular system.* In the section of the midrib note

   a. The position of the *phloem* and *xylem* areas with reference to the leaf surfaces.[8]
   b. The tissues in each as compared with those of the stem bundles.
   In sections of the veinlets note
   c. The tissues that persist.

4. **Draw** a transverse section of the leaf, illustrating all the observed facts.

---

[8] Note how these positions follow from the relative positions of xylem and phloem in the stem.

Mount a **portion of** the epidermis of both surfaces of **the** leaf, **and note**

5. The *epidermal **tissue.***

   a. *Shape* of the cells.
   b. *Hairs* of two kinds.
   c. *Stomata*, comparative abundance on the two sur- faces.
   d. Often dark sheaf-like masses of *raphides*, **torn** out from underlying cells.
   e. Illustrate with drawings.

D. THE FLOWER.

1. The *sepals.* Mount a sepal, examine with a high power, and note

   a. *Venation.*
   b. *Epidermal cells*, their shape and striation.
   c. Presence or absence **of** *stomata*.
   d. Make a drawing, showing these characters.

2. **The *petals*.** Mount a petal, and note

   a. *Venation.*
   b. *Epidermal cells* of the tip and base, their shape and striation.
   c. Presence or absence of *stomata*.
   d. Draw.

3. **The *stamens*.** Make a transverse section through **a** fil- ament and also through an anther of an advanced bud,[9] examine with **a** high power, and note

   a. The structure of the *filament*.
       i.   The epidermis.
       ii. The parenchyma beneath.

---

[9] This **can best be** done by making a section of the entire bud.

     iii. The fibro-vascular bundle, comparing its **tissues** with those in the stem bundles.

b. The two *thecæ*, each **consisting of two valves.**

c. **The wall of the** *valves*, **consisting of**
     i. The epidermis.
     ii. The *endothecium*, an **elastic inner row of** spirally thickened **cells.**

d. Draw a section of **a** theca showing its tissues.

e. The *pollen.* Note
     i. The *surface* of the wall.
     ii. **The two layers of the wall :** the extine colored **and with thin spots ; the** intine thin and colorless.

By careful pressure upon the cover glass, there can be seen
     iii. **The intine** unbroken, but protruding through one of the thin spots in the extine, the true character of the wall becoming thus very obvious.
     iv. **The** minutely granular *contents.*

4. The *pistil.* **Mount a** slice from the surface **of the** stigma and also a transverse section of the ovary, both cleared with potash, and note

a. **The** *stigma.*
     i. Its surface, with pollen tubes sometimes penetrating **it.**
     ii. Draw.

b. The *ovary.*
     i. **The** epidermal cells.
     ii. The character of the mesophyll.
     iii. The fibro-vascular bundles, **their** position and tissues.
     iv. **The structure of the placentæ.**

v.   The structure of the false partition.

vi.   **Draw.**

c.   The *ovules.*

i.   In favorable sections the pollen tubes may be seen entering the ovules. These are easily recognized, as the tube breaks off some dis-tance from the micropyle.

ii.   The fibro-vascular bundle of the funiculus terminating in the ovule.

iii.   **The** two integuments, distinct **from each other** beyond the bend.

iv.   The nucellus, containing a large **cavity, the** *embryo sac,* **which** follows the curve of the ovule. Within the embryo sac

v.   The *embryo,* in various stages of development.

vi.   **Draw, showing all** the above **facts.**

The following phases in the development of the embryo can not be seen in alcoholic specimens, **but may** readily be traced in fresh ones by the **use of** potash as a clearing **agent.**

d.   The *embryo.* Mount some cleared ovules from **an** advanced but unopened bud, press slightly upon **the cover** glass, **and** note

i.   The large curved *embryo sac.*

ii.   In the end of the sac nearest the micropyle, a roundish or oblong cell, the *oosphere.*

iii.   **At** the opposite end **of the sac, a mass of** cells projecting into it.

Mount **ovules from an open** flower, **treat as** before, **note**

iv.   In place of the oosphere a chain of cells, the pro-embryo, with the basal **cell** usually much swollen and **with a group of** cells at the free end of the **chain,** the forming *embryo.*[10]

---

[10] The endosperm, which develops rapidly in angiosperms after fertili-zation, is too transient in this case to make out satisfactorily.

From this point the development of the embryo may be traced with greater or less particularity, by examining ovules in various stages of advancement, until the following condition is seen in seeds from a young pod :

    v.   The pro-embryo has disappeared.

    vi.   The embryo nearly fills the embryo sac, the *cotyledons* beginning almost exactly at the bend.

    vii.   Make drawings illustrating this development.

E. THE FRUIT. Make a tranverse section through as old a fruit as possible, clear with potash, and note

1. The nature of the epidermis, mesophyll, fibro-vascular bundles, placentæ and partition, compared with that studied in the ovary.

2. The *seed*. In transverse sections of seeds note

    a.   The *testa*, its color and structure.

    b.   The *thin-walled tissue* filled with food material.

    c.   The *cotyledons*, the nature of their tissues as compared with those of the leaf.

    d.   The *caulicle*, its structure and tissues as compared with those of the stem. Draw.

    e.   Draw a complete section of the seed, filling in enough of the tissues to indicate their character.

ANNOTATIONS.

Capsella very well presents in a compact form the salient features of a dicotyledon. The paired cotyledons, net-veined leaves, four-parted flowers, and continuous fibro-vascular zone of the stem, all mark it as a member of this highest group.

The **primary** root continues the plant axis below the **surface of** the ground in the form of a tap root, and thus enables the plant to take a deep and firm hold upon the soil. Such primary roots are best developed in dicotyledons and gymnosperms, remaining small in monocotyledons and pteridophytes.

The foliage, instead **of being** somewhat evenly distributed **along the stem and its** branches, is largely collected at the surface of the ground in a cluster **of so-called root-leaves. The toothed and** lobed outline **of the** leaves **with** reticulated venation is quite characteristic of dicotyledons. In Trillium (an anomalous monocotyledon in this regard) there was presented the palmate **type** of net-veined leaves, while in Capsella we find **the pinnate** type, tending to **narrower** and longer **leaf** forms.

An exceptional **feature of** Capsella **(and other** *Cruciferæ*) is the **entire** suppression of bracts in **the flower cluster,** giving **the** pedicels (branches) **the** appearance **of** originating from the main axis without subtending leaves.

The structure of **the** flower **is not** typical of dicotyledons, **in** which the type would **be better** expressed **by an** arrangement like that **of** Trillium, after substituting **five** for three **as the type number. As a** member of the *Cruciferæ,* **however,** Capsella has two whorls **of** two sepals **each, the lower** (outer) being median **(in the** plane of the axis) **and the** inner lateral; **one whorl of** four petals, alternating **with** the four sepals; **two** whorls of stamens, the outer **and** shorter pair **lateral, the** inner and longer set composed **of four** stamens, arranged **in** axial pairs(tetradynamous); **and** one whorl of two carpels laterally placed. There has been much discussion concerning the cruciferous flower, chiefly as to its six stamens and single whorl of four petals. The most natural explanation seems to be that which makes two the type number throughout, the inner whorl of stamens and the single whorl of petals each becoming four by chorisis.[11] The morphological significance of the small glands among the stamens at the base of the ovary is uncertain.[12]

The bi-carpellary ovary becomes two-celled by a membranous outgrowth connecting the two opposite parietal placentæ. This outgrowth, not being a usual part of the carpels, is considered a false or spurious partition. When the fruit (a silicle) opens, the two valves split away from this false partition, to which the placentæ and hence the seeds remain attached.

No part of vascular plants has so constant a character as the root. The root-cap and root-hairs, most characteristic root structures, are much alike in all cases. The primary arrangement of the tissues in pteridophytes, gymnosperms, monocotyledons, and dicotyledons is upon the same plan throughout. The original number of xylem and phloem masses is quite limited in dicotyledons, ranging from two of each (binary, as in Capsella) to eight, but is not constant; while in monocotyledons it is generally larger. In dicotyledons and gymnosperms the root increases in thickness by secondary growth which eventually produces great

---

[11] Gray, Struct. Bot., p. 206, with reference to the views of Eichler, Kunth, Henslow, and others ; Strasburger, Bot Pract , p. 587 , Eichler, Flora, 1865, p. 497, and 1869, p. 97 (both with plates) ; Blüthendiagramme, ii, p. 200, where the literature is cited.

[12] Cf. Hildebrand, Prings. Jahrb., xii, p. 10 ; Müller, ibid, p 161.

changes in the primary structure. Certain of the delicate parenchyma cells lying between the xylem and
phloem elements undergo repeated division, producing
wood and bast tissue. The layer of cambium cells thus
begun on either side of the original plate of xylem soon
unites with its neighbor at the ends, and forms a closed
cambium ring. This ring has the properties of the
cambium layer of the stem, as in Pinus, and by means
of it the root is enabled to increase in thickness
to any extent. It does not, however, as in the
stem, produce its phloem exclusively on the outside and
xylem on the inside of the ring, but they lie side by
side in radiating lines, the number of these lines
increasing with the increase in circumference.[13]

The fundamental system in the stem of dicotyledons is much more differentiated than is usual in monocotyledons. It is divided into an inner and outer
region by the fibro-vascular system, in the latter of
which various tissues may be developed, such as col-
lenchyma, fibrous tissue, etc. In the case of Capsella
the principal modification of the parenchyma of the
fundamental system is the development of the abundant fibrous tissue (sometimes referred to sclerenchyma), which embraces the xylem of the bundles
and arches between the phloem areas. In the fibro-
vascular system the chief characters of the dicotyledonous stem appear. The wedge-shaped bundles are
not scattered through the fundamental tissue, but
are arranged in a zone concentric with the surface of

---

[13] On the secondary thickening of roots see DeBary, Comp. Anat , p.
473 ; Goodale, Physiol. Bot., p. 113 ; VanTieghem, Ann. Sci. Nat., sér.
5, xiii, p. 185.

the stem, and inclosing the inner region of the fundamental tissue, the pith. The parenchyma rays
(medullary rays) left between the bundles may be
broad or narrow. The arrangement and course of the
bundles depend largely upon the position of the leaves.
From each leaf one or more bundles enter the stem
and passing downward finally become part of the fibro-
vascular zone. Transverse sections of the stem often
cut across bundles midway in their course from the
leaf to the vascular ring, and they then appear as if
belonging to the cortex. The bundles are collateral,
with a cambium layer between the xylem and phloem,
forming the characteristic open bundle of dicotyledons. In Capsella a bundle-sheath arches over each
bundle, and frequently becomes continuous around
the entire fibro-vascular zone.[14] In the xylem the
spiral and annular vessels are the oldest and most
centrally placed, the dotted ducts, the largest elements
of the xylem, occurring nearest the phloem [15]

The leaf shows the general dicotyledonous characters of more contorted epidermal cells and more numerous and smaller stomata. The fibro-vascular bundles are like those of the stem, tracheides replacing
other vascular elements in the ultimate ramifications.
Capsella is so favorable for the study of the
development of the embryo, that this very important subject has been deferred until now. It has
already been seen how the asexually produced pollen
spore (microspore), after falling upon the papillated

---

[14] Pointed out by Kamienski, in DeBary's Compar. Anat., p. 415.

[15] For stem structure see Prantl and Vines, Text-book, p. 47 ; Bessey,
Bot., p. 438 ; Goodale, Physiol. Bot., p. 119.

surface of the stigma, develops a pollen-tube and penetrates the tissues of the style. The rate of descent of
the pollen-tube is quite various in different plants. In
the style and walls of the ovary there is usually a
region of least resistance to penetration, furnished by
the delicate "conducting tissue," or the style is
frequently tubular (as in Viola). In Capsella, very
soon after pollination, an abundance of pollen-tubes is
found in the ovarian cavity. Some of them may be
seen to have entered the micropyles of the ovules and
penetrated to the nucellus.

The preparation of the ovule for fertilization has
been the development, at the apex of the nucellus, of
the embryo sac (macrospore), at the micropylar end of
which lies the oosphere (embryonal vesicle), accompa-
nied usually by two similar masses, the synergidæ. At
the base of the embryo sac appear three or more free
cells, the antipodal cells[16] of Hofmeister. The six
cells which differentiate into the antipodal cells,
oosphere and synergidæ, constitute a very rudimentary
prothallium,[17] which is far more reduced than in gymnosperms, but corresponds to the primary endosperm
of these plants. The endosperm (of most text-books),
more properly secondary endosperm, is produced by
cell-formation around the nuclei arising from division
of the definitive nucleus of the embryo sac.[18] When

---

[16] Strasburger, Bot. Pract., p. 522, et seq.; Prantl and Vines, Text-
book, p. 205.

[17] Sachs, Text-book, 2nd Eng. ed., p. 582, where a fuller account
of the changes preliminary to fertilization in angiosperms may be
found.

[18] Sachs, Text-book, 2nd Eng. ed., p. 585.

fertilization[19] has taken place a membrane is developed about the oosphere, making it a sexual spore.

By divisions[20] in one plane the oospore at once extends toward the interior of the ovule as a chain of cells, the suspensor or pro-embryo, the basal cell of which becomes large and bladder-like. The apical cell at the free end of the suspensor, by repeated division in several planes, forms a cell mass, which presently assumes the form of the embryo.[21] The ovule after various changes of minor importance in this connection becomes at last a ripe seed.

---

[19] For an account of the nuclei of the pollen spore and oosphere, and their union in the fertilizing act, see Strasburger, Neue Untersuchungen.

[20] For methods of cell division in the developing embryo of Capsella (with figures) see Bessey, Bot., p. 424 ; Westermaier, Die ersten Zelltheilungen im Embryo von Capsella, Flora, 1876, p. 483.

[21] For further description of the development of the embryo see Gray, Struct. Bot., p. 283 ; Prantl and Vines, Text-book, p. 204 : Bessey, Bot., p. 423 ; Sachs, Text-book, 2nd Eng. ed., p. 585.

GLOSSARY.

Ab-stric'e-tion (ab, *off*; stringo, *I tie*). Partial or complete separation by contraction.

A-nat'ro-pous (ἀνά, *up*; τρέπω, *I turn*). Said of an inverted ovule or seed which has the rhaphe extending its whole length.

An-droe'ci-um (ἀνήρ, *a male*; οἶκος, *a house*). The stamens of a flower collectively.

A'n-nu-lus (annulus, *a small ring*). The elastic ring of cells around the sporangium in ferns.

A'n-ther (ἀνθηρός, *flowery*). The pollen-bearing part of the stamen.

An-ther-i'd-i-um, pl. antheridia (anther; εἶδος, *form*). The male organ of the lower groups, analogous to but not homologous with the anther of phanerogams.

A'n-ther-o-zoids (anther; ζῶον, *an animal*; εἶδος, *form*). The male reproductive bodies developed in antheridia.

A'n-tho-tax-y (ἄνθος, *a flower*; τάξις, *arrangement*). The arrangement of flowers in a cluster; inflorescence.

An-ti'p-o-dal (ἀντί, *over against*; πούς, *a foot*). Said of a group of cells at the end of the embryo-sac furthest from the micropyle.

A'p-i-cal (apex, *the top*). At the apex or tip.

A-po'ph-y-sis (ἀπό, *from*; φύσις, *nature*). In mosses, an enlargement of the pedicel at the base of the capsule.

Arch-e-go'n-i-um, pl. archegonia (ἀρχή, *beginning*; γονή, *offspring*).

The female organ of bryophytes and pteridophytes.

A-re'o-la, pl. areolæ (areola, *a small open space*). The spaces in a reticulated surface, as in the thallus of Marchantia.

A's-co-spores. The spores formed in an ascus.

A's-cus, pl. asci (ἀσκός, *a sac*). The spore sac of a large group of carpophytes.

A'x-i-al. Relating or belonging to the axis.

A'x-il (axilla, *the arm-pit*). The point just above the attachment of a leaf to the stem.

A'x-is (axis, *an axle-tree*). The central part or longitudinal support on which organs or parts are arranged.

Bast (bass). In general, the phloëm region of a fibro-vascular bundle; or, specifically, the fibers of the phloëm.

Bract (bractea, *a thin plate*). The more or less modified leaves of a flower cluster.

Bry-o'ph-y-ta (βρύον, *moss*; φυτόν, *a plant*). A primary division of plants, named from its principal group, the mosses. *Bry'-o-phyte* is the English equivalent.

Bul'li-form (bulla, *a swelling*). Said of enlarged or swollen cells.

Ca'l-lus (callus, *a callosity*). A hardened or thickened place; technically used of the thickening mass in a sieve-plate, usually appearing as a layer on each side of the plate.

Ca-ly'p-tra (καλύπτρα, *a cover*). In mosses, the hood which covers the capsule.

Ca'-lyx (calyx, *a cup*). The outer envelope of a flower, composed of sepals.

Ca'm-bi-form. Resembling cambium.

Ca'm-bi-um (cambio, *I exchange*). The meristem cells of an open fibrovascular bundle, lying between the phloëm and xylem, which retain the power of division.

Cam-py-lo't-ro-pous (καμπή, *bending;* τρέπω, *I turn*). Said of an ovule or seed which becomes curved in its growth so as to be inverted.

Ca'p-sule (capsula, *a small box*). A dry dehiscent seed-vessel (formed of more than one carpel) ; or a similar spore-vessel.

Ca'r-pel (καρπός, *fruit*). The constituent leaf of a pistil · hence either a simple pistil, or one of the parts of a compound pistil.

Ca'r-pel-la-ry. Relating to a carpel.

Car-po-go'-ni-um, pl. carpogonia (καρπός, *fruit;* γονή, *offspring*). The female organ of carpophytes.

Ca'r-po-phyll (καρπός, *fruit;* φύλλον, *a leaf*). The carpellary leaf.

Car-po'ph-y-ta (καρπός, *fruit;* φυτόν, *a plant*). A primary division of plants, named from the sporocarp, or spore-vessel, which is the result of fertilization. *Ca'r-po-phyte* is the English equivalent.

Car-y-o'p-sis (κάρυον, *a nut;* ὄψις, *an appearance*). A grain ; the seed-like fruit of grasses.

Cau'-li-cle (cauliculus, *a small stem*). The initial stem in an embryo.

Cell (cella, *a cell*). The anatomical unit of plant-structure.

Ce'l-lu-lose (cellulosus, *pertaining to a cell*). The primary substance of the cell-wall.

Chaff. Small dry scales.

Cha-la'-za (χάλαζα, *that which is let loose*). The part of an ovule where integuments and nucellus are confluent.

Chlo'-ro-phyll (χλωρός, *greenish-yellow;* φύλλον, *a leaf*). The green coloring matter of plants.

Cho'r-i-sis (χώρισις, *a separating*). Longitudinal separation into two or more similar parts.

Ci'l-i-um, pl. cilia (cilium, *an eyelash*). Marginal hairs ; motile protoplasmic filaments, as those of antherozoids.

Closed bundle. A fibro-vascular bundle containing no cambium.

Col-la't-er-al (con, *together;* latus, *a side*). Side by side ; used of a fibrovascular bundle in which the xylem and phloëm are side by side in a radial direction.

Col-u-me'l-la (columella, *a small column*). The persistent axis of certain spore-cases, as in mosses.

Con-ce'n-tric (con, *together;* centrum, *the center*). Technically used of a fibro-vascular bundle whose tissues are arranged concentrically.

Co-nl' di-o-phore (conidia ; φέρω, *I carry*). The stalk upon which conidia are borne.

Co-ni'-di-um (gonidium), pl. conidia (γονή, *offspring;* εἶδος, *form*). The asexual spores of certain groups.

Con-ju-ga'-tion (conjugatus, *joined together, paired*). The sexual union of similar cells, as in zygophytes.

Con-ne'ct-ive (connecto, *I connect*). The portion of the stamen connecting the thecæ.

Co-ro'l-la (corolla, *a small crown*). The inner envelope of a flower, within the calyx, and composed of petals.

Cor-pu's-cu-lum, pl. corpuscula (corpusculum, *a little body*). The archegonium-like structures in the ovule of gymnosperms.

Co'r-tex (cortex, *the bark*). The rind or bark.

Co'r-ti-cal. Relating to the cortex or bark.

Cot-y-le'-don (κοτυληδών, *a cup-shaped cavity*). A primary embryo-leaf borne by the caulicle.

Cu'-pule (cupula, *a little tub*). The gemma-cup of liverworts.

Cu'-ti-cle (cuticula, *the skin*). The outermost film or pellicle of the epidermis, differing chemically from the remainder of the cell-wall.

Der-ma'-to-gen (δέρμα, *skin*; γεννάω, *I produce*). The layer of nascent epidermis in the meristem of growing points.

Di-cho't-o-mous (δίχα, *in two*; τέμνω, *I cut*). Forking regularly by pairs.

Di-cot-y-le'd-o-nous (δίς, *double*; cotyledon). Having two cotyledons, or seed-leaves.

Di-œ'-cious (δίς, *double*; οἶκος, *a house*). Having the two sex-organs borne by distinct individuals.

E-la'-ter (ἐλατήρ, *one that expels*). Spirally thickened cells within the sporogonia of some liverworts, which assist in expelling the spores.

Em-bry-o (ἔμβρυον, *fœtus*, or *embryo*). The young plantlet within the seed.

Embryo-sac. The cavity, within the nucellus, in which the embryo develops.

En-do-de'rm-is (ἔνδον, *within*; δέρμα, *the skin*). The layer of cells inclosing the fibro-vascular bundle; the bundle sheath.

En-do'g-e-nous (ἔνδον, *within*; γεννάω, *I produce*). Originating from internal tissues, and penetrating the outer ones.

En-do-sperm (ἔνδον, *within*; σπέρμα, *the seed*). A parenchymatous tissue developed within the embryo-sac.

En-do-spore (ἔνδον, *within*; spore). The inner layer of a spore-wall.

En-do-the'-ci-um (ἔνδον, *within*; theca). The inner wall of the theca.

Ep-i-de'rm-is (ἐπί, *upon*; δέρμα, *the skin*). The outermost layer of special cells covering plant-surfaces.

E'p-i-phragm (ἐπί, *upon*; φράγμα, *a protection*). In mosses, a membrane covering the orifice of the capsule.

Ex-o'g-e-nous (ἔξω, *outside*; γεννάω, *I produce*). Originating from outer layers of tissue.

E'x-o-spore (ἔξω, *outside*; spore). The outer layer of a spore-wall.

E'x-tine (exter, *on the outside*). The outer coat of a pollen-spore.

Fi'-ber (fibra, *a fiber*). A long and slender, thick-walled cell.

Fi'-brous. Composed of fibers.

Fi-bro-va's-cu-lar (fibra, *a fiber*; vasculum, *a small vessel*). Composed of fibers and vessels; fibro-vascular bundles are the strands which make up the framework of the higher plants.

Fi'l-a-ment (filum, *a thread*). The stalk of the stamen, supporting the anther; also the individual threads of algæ or fungi.

Flowering glume. In grasses, the bract which subtends each flower, sometimes called lower palet.

Frond (frons, *a leaf*). A name given to the leaves of ferns.

Fundamental tissue. That outside the fibro-vascular bundles and inclosed by the epidermis, but not included in either.

Fu-ni'c-u-lus (funiculus, *a slender rope*). The stalk of an ovule or seed.

Gem'-ma, pl. gemmæ (gemma, *a bud*). In bryophytes, many-celled bodies for asexual propagation.

Glau'-cous (γλαυκός, *pale green, gray*). Whitened with a bloom, like that on a cabbage-leaf.

Glume (gluma, *a husk*). A chaff-like bract belonging to the inflorescence of grasses; the outer glumes subtend the spikelet; the flowering glume is the bract of the flower.

Glu'ten (gluten, *glue*). A general term for the glue-like products of plants, especially of seeds.

Grain. A seed-like fruit, like those of grasses, with pericarp adnate to the seed; also any small rounded body, as of starch or chlorophyll.

Growing point (*punctum vegeta-*

tionis). The group of meristem cells at the growing tip of an organ, from which the various tissues arise.

Gy-nœ'-ci-um (γυνή, *a female;* οἶκος, *a house*). The pistil, or collective pistils, of a flower.

Haus-to'-ri-a (haustor, *a drinker*). The absorbing organs of certain parasitic plants.

Her-ma'ph-ro-dite (ἑρμαφρόδιτος, *one who is both male and female*). Having both kinds of sexual organs borne together on the same axis.

Host. The plant upon which parasitic plants [or organisms] develop, and from which they derive their nourishment.

Hy-gro-sco'p-ic (ὑγρος, *wet;* σκοπέω, *I look out for*). Having an avidity for water.

Hy-me'n-i-um (ὑμήν, *a membrane*). In fungi, a surface layer of vertical filaments containing or bearing spores.

Hy'-pha, pl. hyphæ (ὑφή, *a web*). The slender vegetative filaments of fungi which may or may not be woven into a mat (mycelium), or a plant body.

Hy-po-de'r-ma (ὑπό, *under;* δέρμα, *the skin*). The thick-walled tissues beneath the epidermis, which serve to strengthen it, but do not belong to the fibro-vascular bundle.

In-cu'm-bent (incumbo, *I lean upon*). Leaning or resting upon; of cotyledons, when the radicle is against the back of one; of anthers, when they lie against the inner face of the filament.

In-du'-si-um, pl. **indusia** (indusium, *clothing*). In ferns, a cellular outgrowth of the leaf covering the clusters of sporangia (sori).

In-flor-e'n-cence (infloresco, *I blossom*). The arrangement of flowers; or the flowering portion of a plant.

In-ter-ce'l-lu-lar. Between or among the cells.

In'-ter-node (inter, *between;* nodus, *a joint*). The part of a stem between two nodes.

In-tine (inter, *on the inside*). The inner coat of a pollen-spore.

La'm-i-na (lamina, *a layer*). The blade, or expanded part, of a leaf.

Leaf-trace. The fibro-vascular bundles from the leaf which descend into the stem, and sooner or later become blended with its fibro-vascular system.

Li'g-ule (ligula, *a small tongue*). In grasses, a thin appendage at the junction of leaf-blade and sheath.

Lo'd-i-cule (lodicula, *a small coverlet*). A small scale in the flower of grasses.

Ma'c-ro-spore (μακρός, *large;* spore). The larger spore of the two kinds produced by certain pteridophytes.

Me'd-ul-la-ry (medulla, *pith*). Relating to the pith; medullary rays are the pith-rays which pass outward to the bark between the fibro-vascular bundles.

Mer-i's-tem (μερίζω, *I divide*). Tissues in a nascent or differentiating state.

Me's-o-phyll (μέσος, *middle;* φύλλον, *a leaf*). The green or soft tissue of a leaf, supported by the framework and exclusive of the epidermis, called by the older botanists parenchyma.

Mi'-cro-pyle (μικρός, *small;* πύλη, *a gate*). The opening left by the integuments of the ovule, and which leads to the nucellus.

Mi'-cro-spore (μικρός, *small;* spore). The smaller spore of the two kinds produced by certain pteridophytes.

Mi'd-rib. The central or main rib of a leaf or thallus.

Mon-o-po'-di-al (μόνος, *single;* πούς, *a foot*). Said of a stem consisting of a single and continuous axis (footstalk).

My-ce'-li-um (μύκης, *a mushroom;* λίς, *cloth*). The filamentous vegetative growth of fungi, composed of hyphæ.

Naked. Wanting some usual covering.

Nec-ta-ry (nectarium, *a depository for nectar*). The place or appendage in which nectar is secreted.

Nerve (nervus, *a nerve*). **A simple** vein or rib.

Node (nodus, *a joint*). That part of a stem which normally bears leaves.

Nu-cel-lus (nucella, *a little kernel*). The mass of the ovule within the integuments, sometimes called the nucleus.

Nu-cle'-o-lus (diminutive of nucleus). The sharply defined point often seen in the nucleus.

Nu'-cle-us (nucleus, *a kernel*). The usually roundish mass found in the protoplasm of most active cells, and differing from the rest of the protoplasm in its greater density.

O-o-go'-ni-um, pl. **oogonia** (ᾠόν, *an egg*; γονή, *offspring*). The female organ of oöphytes.

O-o'-ph y-ta (ᾠόν, *an egg*; φυτόν, *a plant*). A primary division of plants, named, from the mode of reproduction, the egg-spore plants. *O'-o-phyte* is the English equivalent.

O'-o-sphere (ᾠόν, *an egg*; σφαῖρα, *a sphere*). The naked female egg-cell; the mass of protoplasm prepared for fertilization.

O'-o-spore (ᾠόν, *an egg*; spore). In general, the egg-cell after fertilization, and surrounded by a cell-wall; also specially applied to the spore formed in an oögonium.

Open bundle. A fibro-vascular bundle which contains cambium.

O per'-cu-lum, pl. **opercula** (operculum, *a cover*). In mosses, the terminal lid of the capsule.

O'-va-ry (ovarius, *an egg-keeper*). That part of the pistil which contains the ovules.

O'-vule (diminutive of ovum, *an egg*). The body which becomes a seed after fertilization.

Pa'-let (palea, *chaff*). In grasses, the inner bract of the flower.

Pal-i-sade cells. The elongated parenchyma cells of a leaf, which stand at right angles to its surface, and are usually confined to the upper part.

Pal-mate (palma, *the hand*). Radiating like the fingers; said of the veins or divisions of some leaves.

Pan-i-cle (panicula, *a tuft*). **A loose** and irregularly branching **flower**-cluster, as **in many** grasses.

Par-a'ph-y-sis, pl. **paraphyses** (παρά, *beside*; φύσις, *nature*). Sterile bodies, usually hairs, which are found mingled with the reproductive organs of various cryptogams.

Pa-ren-chy-ma (παρεγχέω, *I pour in beside*). Ordinary or typical cellular tissue, *i.e.* of thin-walled, nearly isodiametric cells.

Par-the-no-ge'n-e-sis (παρθένος, *a virgin*; γένεσις, *generation*). Commonly applied to the production of **seed** without fertilization; but, strict**ly, the** formation **of a** sexual spore **without fertilization.**

Ped-i-cel (pediculus, *a little foot*). **The stalk upon which an organ is** borne.

Pe-du'n-cle (pedunculus, *a little foot*). The general flower-stalk.

Pe'r-i-anth (περί, *around*; ἄνθος, *a flower*). The floral envelopes, or leaves of a flower, taken collectively; and an analogous envelope of the sporogonia of certain liverworts.

Pe'r-i-blem (περίβλημα, *a covering*). A name given to that part of the meristem at the growing point of the plant-axis, which lies just beneath the epidermis and develops into the cortex.

Per-i-ca'm-bi-um (περί, **around**; cambium). In roots, the **external** layer of the fibro-vascular **cylinder.**

Per-i-chæ'-ti-um, pl. **perichætia** (περί, **around**; χαίτη, *hair*, or *leaf*). **In bryophytes, the** leaves or leaf-like parts which envelop the clusters of sex-organs, forming in some cases the so-called flower.

Pe'r·i·stome (περί, *around ;* στόμα, *a mouth*). In mosses, usually bristle-like or tooth-like structures surrounding the orifice of the capsule.

Per·i the'·ci·um, pl. **perithecia** (περί, *around ;* θήκη, *a case*). The spore-vessel of certain carpophytes, containing the spore-sacs (asci).

Pe't·al (πέταλον, *a leaf*). A corolla leaf.

Pe't·i·ole (petiolus, *a little foot*). The stalk of a leaf.

Phan·e·ro·ga'·mi·a (φανερός, *evident ;* γάμος, *marriage*). A primary division (the highest) of plants, named from their mode of reproduction, the seed-producing plants. *Pha'n·e·ro·gam* is the English equivalent.

Phlo'·em (φλοιός, *the inner bark*). The bark or bast portion of a fibro-vascular bundle.

Phy·co·cy'·an·ine (φῦκος, *sea-weed ;* κύανος, *dark blue*). A bluish coloring matter extracted by water from certain algæ.

Phy'l·lo·tax·y (φύλλον, *a leaf ;* τάξις, *arrangement*). Leaf-arrangement.

Pi'n·na, pl. **pinnæ** (pinna, *a feather*). One of the primary divisions of a pinnate leaf, as in ferns.

Pi'n·nule (pinnula, *a little feather*). One of the divisions of a pinna.

Pis·til (pistillum, *a pestle*). The female organ in phanerogams.

Pit. A thin place, or pit-like depression, left in the thickening of a cell-wall.

Pla·cen'·ta, pl. placentæ (placenta, *a cake*). That portion of the ovary which bears the ovules.

Ple'·rome (πλήρωμα, *that which fills*). A name given to that part of the meristem near the growing points of the plant-axis, which forms a central shaft or cylinder and develops into the axial tissues.

Plu'·mule (plumula, *a small, soft feather*). The terminal bud of the embryo above the cotyledons.

Pod. A dry, several-seeded, dehiscent fruit; or a similar spore-case.

Pol·len (pollen, *fine flour*). **The** spores developed in the anther.

Pol·lin·a'·tion. The transfer of pollen to its stigma.

Pro·embryo (pro, *before ;* embryo). In phanerogams, the chain of cells (suspensor) formed after fertilization, and from the lower end of which the embryo develops.

Pro·thal'·li·um, pl. **prothallia** (pro, *before ;* thallus, *a young shoot*). In pteridophytes, the small usually short-lived plant which develops from the spore, and bears the sex-organs.

Pro·to·ne'·ma, pl. protone'mata (πρῶτος, *first ;* ἦμα, *that which is sent out*). In mosses, the filamentous growth which is produced by the spores, and from which the leafy moss plant is developed.

Pro·to'ph·y·ta (πρῶτος, *the first ;* φυτόν, *a plant*). A primary division of plants, named from the fact that they include the lowest known plants. *Pro'·to·phyte* is the English equivalent.

Pro'·to·plasm (πρῶτος, *first ;* πλάσμα, *that which has been formed*). That substance in living cells, of varying consistency, which is the seat of all vital phenomena.

Pte'r·i·doid (πτέρις, *a fern ;* εἶδος, *form*). Fern like.

Pter·i·do'ph·y·ta (πτέρις, *a fern ;* φυτόν, *a plant*). A primary division of plants, named from its principal group, the ferns. *Pte·ri'd·o·phyte* is the English equivalent.

Py'r·e·noid (πυρήν, *kernel ;* εἶδος, *form*). Minute colorless bodies imbedded in the chlorophyll structures of some lower plants.

Ra'ph·i·des (ῥαφίς, *a needle ;* εἶδος, *form*). Needle-like plant-crystals.

Re·ce'p·ta·cle (receptaculum, *a receptacle*). That portion of an axis or pedicel (usually broadened) which forms a common support for a cluster of organs, in most cases sex-organs.

Re·ti'c·u·la·ted (reticulatus, *net-like*). Having a net-like appearance.

Rha'-chis (ῥάχις, *the backbone*). The axis of a compound leaf, or of a spike.

Rha'-phe (ῥαφή, *a seam*). In an anatropous ovule, the ridge which connects the chalaza with the hilum.

Rhi'-zoid (ῥίζα, *a root; εἶδος, form*). Root-like; a name applied to the root-like hairs found in bryophytes and pteridophytes.

Rhi'-zo-tax-y (ῥίζα, *a root; τάξις, arrangement*). Root-arrangement.

Root-stock. A horizontal, more or less thickened, root-like stem, either on the ground or underground.

Sca-la'r-i-form (scalaria, *a ladder;* forma, *form*). A name applied to ducts with pits horizontally elongated and so placed that the intervening thickening ridges appear like the rounds of a ladder.

Scale (scala, *a flight of steps*). Any thin scarious body, as a degenerated leaf, or flat trichome.

Scle-re'n-chy-ma (σκληρός, *hard;* ἔγχυμα, *an infusion*). A tissue belonging to the fundamental system and composed of cells that are thick-walled, often excessively so.

Scu-te'l-lum (scutella, *a small disk*). The disk-like or shield-like cotyledon of grasses.

Seed. The fertilized and matured ovule.

Se'p-al (from the modernized word σέπαλον, *a sepal*). A calyx leaf.

Se'-ta, pl. setæ (seta, *a bristle*). A bristle, or bristle-shaped body; in mosses, the stalk of the capsule.

Sheath. A thin enveloping part, as of a filament, leaf, or resin-duct.

Sieve-cells. Cells belonging to the phloëm, and characterized by the presence of circumscribed and perforated panels in the walls; the panels are *sieve-plates*, and the perforations *sieve-pores*.

So'-rus, pl. sori (σωρός, *a heap*). In ferns, the groups of sporangia, constituting the so-called "fruit-dots;" in parasitic fungi, well-defined groups

of spores, breaking through the epidermis of the host.

Spike (spica, *an ear of corn*). A flower-cluster, having its flowers sessile on an elongated axis.

Spi'ke-let (diminutive of spike). A secondary spike; in grasses, the ultimate flower-cluster, consisting of one or more flowers subtended by a common pair of glumes.

Spo-ra'n-gi-um, pl. sporangia (spore; ἄγγος, *a vessel*). The spore-vessel; applied to ferns and certain lower groups.

Spore (σπορά, *seed*). Originally used as the analogue of seed in flowerless plants; now applied to any one-celled or few-celled body which is separated from the parent for the purpose of reproduction, whether sexually or asexually produced; the different methods of its production are indicated by suitable prefixes.

Spo-ro-go'-ni-um, pl. sporogonia (spore; γονή, *offspring*). The whole structure of the spore-bearing stage of bryophytes.

Sta'-men (στήμων, *the warp or thread of cloth*). The male organ in phanerogams.

Stig'-ma (στίγμα, *a spot*, or *mark*). The surface of a pistil without epidermis which receives the pollen.

Stig-ma'-tic. Relating to the stigma, or stigma-like.

Sto'-ma, pl. sto'mata (στόμα, *a mouth*). Epidermal structures which serve for facilitating gaseous interchanges with the external air, often called "breathing-pores."

Stro'-phi-ole (strophiolum, *a small wreath*). An appendage at the hilum of certain seeds.

Style (στῦλος, *a pillar*). The usually attenuated portion of the pistil which bears the stigma.

Sus-pe'n-sor (suspendo, *I hang*). See Pro-embryo.

Syn-e'r-gi-dæ, or Synergides (συνεργέω, *I work together*). The two nucleated bodies which accompany the oösphere in the embryo-sac, and together with it form the egg-apparatus.

Te's-ta (testa, *a shell*). The outer seed-coat.

Tet-ra-dy'n-a-mous (τετράς, *four;* δύναμις, *strength*). Said of an androecium in which there are four long and two shorter stamens.

Tha'l-loid (thallus; εἶδος, *form*). Thallus-like.

Tha'l-lus (θαλλός, *a young shoot*). The body of lower plants, which exhibits no differentiation of stem, leaf, and root.

The' ca, pl. **thecæ** (θήκη, *a case*). The "anther-cell," that is, the case containing pollen; sometimes used of other spore-cases.

Tra'cheary tissue. A general name given to the vessels and ducts found in fibro vascular bundles.

Tra'-che-ides (τραχύς, *rough;* εἶδος, *form*). Tracheary cells that are closed throughout.

Tri'-chome (θρίξ, *hair*). A general name for a slender outgrowth from the epidermis, usually arising from a single cell.

Tur-gi'd-i-ty (turgidus, *swollen*). The normal swollen condition of cells which results from the avidity of protoplasm for water.

Vein (vena, *a vein*). The fibro-vascular bundle of leaves or any flat organ.

Ve-na'-tion (vena, *a vein*). The mode of vein distribution.

Xy'-lem (ξύλον, *wood*). The wood (inner) portion of the fibro-vascular bundle.

Zo'-o-spore (ζῷον, *an animal;* spore). A free-moving spore.

Zy-go-mo'r-phic (ζυγόν, *a yoke;* μορφή, *form*). Said of a flower which can be bisected by only one plane into similar halves.

Zy-go'ph-y-ta (ζυγόν, *a yoke;* φυτόν, *a plant*). A primary division of plants, named from their mode of reproduction, the sexual spore being produced by conjugation. *Zy'-go-phyte* is the English equivalent.

Zy'-go-spore (ζυγόν, *a yoke;* spore). The spore of zygophytes, formed by conjugation.

INDEX.

Numbers in light type refer to the **laboratory** part, those in heavy type to other parts of the book.

THE AMERICAN SCIENCE SERIES.

The principal objects of the series are to supply the lack—in some subjects very great—of authoritative books whose principles are, so far as practicable, illustrated by familiar American facts, and also to supply the other lack that the advance of Science perennially creates, of text-books which at least do not contradict the latest generalizations. The scheme systematically outlines the field of Science, as the term is usually employed with reference to general education, and includes ADVANCED COURSES for maturer college students, BRIEFER COURSES for beginners in school or college, and ELEMENTARY COURSES for the youngest classes. The Briefer Courses are not mere abridgments of the larger works, but, with perhaps a single exception, are much less technical in style and more elementary in method. While somewhat narrower in range of topics, they give equal emphasis to controlling principles. The following books in this series are already published:

THE HUMAN BODY. By H. NEWELL MARTIN, Professor in the Johns Hopkins University.

Advanced Course. Large 12mo. Pp. 655. $2.75.

Designed to impart the kind and amount of knowledge every educated person should possess of the structure and activities and the conditions of healthy working of the human body. While intelligible to the general reader, it is accurate and sufficiently minute in details to meet the requirements of students who are not making human anatomy and physiology subjects of special advanced study. *The regular editions of the book contain an appendix on Reproduction and Development. Copies without this will be sent when specially ordered.*

From the CHICAGO TRIBUNE: "The reader who follows him through to the end of the book will be better informed on the subject of modern physiology in its general features than most of the medical practitioners who rest on the knowledge gained in comparatively antiquated text-books, and will, if possessed of average good judgment and powers of discrimination, not be in any way confused by statements of dubious questions or conflicting views."

THE HUMAN BODY—*Continued.*

Briefer Course. 12mo. Pp. 364. $1.50.

Aims to make the study of this branch of Natural Science a source of discipline to the observing and reasoning faculties, and not merely to present a set of facts, useful to know, which the pupil is to learn by heart, like the multiplication-table. With this in view, the author attempts to exhibit, so far as is practicable in an elementary treatise, the ascertained facts of Physiology as illustrations of, or deductions from, the two cardinal principles by which it, as a department of modern science, is controlled,—namely, the doctrine of the "Conservation of Energy" and that of the " Physiological Division of Labor." To the same end he also gives simple, practical directions to assist the teacher in demonstrating to the class the fundamental facts of the science. *The book includes a chapter on the action upon the body of stimulants and narcotics.*

From HENRY SEWALL, *Professor of Physiology, University of Michigan :* "The number of poor books meant to serve the purpose of text-books of physiology for schools is so great that it is well to define clearly the needs of such a work : 1. That it shall contain accurate statements of fact. 2. That its facts shall not be too numerous, but chosen so that the important truths are recognized in their true relation. 3. That the language shall be so lucid as to give no excuse for misunderstanding. 4. That the value of the study as a discipline to the reasoning faculties shall be continually kept in view. I know of no elementary text-book which is the superior, if the equal, of Prof. Martin's, as judged by these conditions."

Elementary Course. 12mo. Pp. 261. 90 cts.

A very earnest attempt to present the subject so that children may easily understand it, and, whenever possible, to start with familiar facts and gradually to lead up to less obvious ones. *The action on the body of stimulants and narcotics is fully treated.*

From W. S. PERRY, *Superintendent of Schools, Ann Arbor, Mich. :* "I find in it the same accuracy of statement and scholarly strength that characterize both the larger editions. The large relative space given to hygiene is fully in accord with the latest educational opinion and practice ; while the amount of anatomy and physiology comprised in the compact treatment of these divisions is quite enough for the most practical knowledge of the subject. The handling of alcohol and narcotics is, in my opinion, especially good. The most admirable feature of the book is its fine adaptation to the capacity of younger pupils. The diction is simple and pure, the style clear and direct, and the manner of presentation bright and attractive."

ASTRONOMY. By Simon Newcomb, Professor in the Johns
Hopkins University, and Edward S. Holden, Director of
the Lick Observatory.

Advanced Course. Large 12 mo. Pp. 512. $2.50.

To facilitate its use by students of different grades, the sub-
ject-matter is divided into two classes, distinguished by the size
of the type. The portions in large type form a complete course
for the use of those who desire only such a general knowledge
of the subject as can be acquired without the application of advanced mathematics. The portions in small type comprise ad-
ditions for the use of those students who either desire a more
detailed and precise knowledge of the subject, or who intend to
make astronomy a special study.

From C. A. Young, *Professor in Princeton College:* "I conclude
that it is decidedly superior to anything else in the market on the
same subject and designed for the same purpose."

Briefer Course. 12mo. Pp. 352. $1.40.

Aims to furnish a tolerably complete outline of the astronomy of to-day, in as elementary a shape as will yield satisfactory returns for the learner's time and labor. It has been
abridged from the larger work, not by compressing the same
matter into less space, but by omitting the details of practical
astronomy, thus giving to the descriptive portions a greater
relative prominence.

From The Critic: "The book is in refreshing contrast to the
productions of the professional schoolbook-makers, who, having only
a superficial knowledge of the matter in hand, gather their material,
without sense or discrimination, from all sorts of authorities, and
present as the result an *indigesta moles*, a mass of crudities, not un-
mixed with errors. The student of this book may feel secure as to
the correctness of whatever he finds in it. Facts appear as facts, and
theories and speculations stand for what they are, and are worth."

From W. B. Graves, *Master Scientific Department of Phillips
Academy:* "I have used the Briefer Course of Astronomy during the
past year. It is up to the times, the points are put in a way to interest the student, and the size of the book makes it easy to go over the
subject in the time allotted by our schedule."

From Henry Lefavour, *late Teacher of Astronomy, Williston Semi-
nary:* "The impression which I formed upon first examination, that
it was in very many respects the best elementary text-book on the
subject, has been confirmed by my experience with it in the class-
room."

ZOOLOGY. By A. S. PACKARD, Professor in Brown University.

Advanced Course. Large 12mo. Pp. 719. $3.00.

Designed to be used either in the recitation-room or in the laboratory. It will serve as a guide to the student who, with a desire to get at first-hand a general knowledge of the structure of leading types of life, examines living animals, watches their movements and habits, and finally dissects them. He is presented first with the facts, and led to a thorough knowledge of a few typical forms, then taught to compare these with others, and finally led to the principles or inductions growing out of the facts.

From A. E. VERRILL, *Professor of Zoology in Yale College:* "The general treatment of the subject is good, and the descriptions of structure and the definitions of groups are, for the most part, clear, concise, and not so much overburdened by technical terms as in several other manuals of structural zoology now in use."

Briefer Course. 12mo. Pp. 334. $1.40.

The distinctive characteristic of this book is its use of the *object method.* The author would have the pupils first examine and roughly dissect a fish, in order to attain some notion of vertebrate structure as a basis of comparison. Beginning then with the lowest forms, he leads the pupil through the whole animal kingdom until man is reached. As each of its great divisions comes under observation, he gives detailed instructions for dissecting some one animal as a type of the class, and bases the study of other forms on the knowledge thus obtained.

From HERBERT OSBORN, *Professor of Zoology, Iowa Agricultural College:* "I can gladly recommend it to any one desiring a work of such character. While I strongly insist that students should study animals from the animals themselves,—a point strongly urged by Prof. Packard in his preface,—I also recognize the necessity of a reliable text-book as a guide. As such a guide, and covering the ground it does, I know of nothing better than Packard's."

From D. M. FISK, *Professor of Natural History, Hillsdale College:* "The 'Briefer Courses' of Packard and Martin have been adopted, and for these reasons: 1. *They are brief;* the lessened mechanical labor of mastering a text leaves time for more observation and for comparison of authorities. 2. *They are clear;* the work of cutting away needless nomenclature has been done with skill. 3. *They are authoritative;* serious students can have confidence in even brief and dogmatic statements, knowing they come from a master, and not from a mere compiler. 4. *They are fresh;* fossils are good in their places, but a fossil text-book in science is a fraud on youth."

ZOOLOGY—*Continued.*

Elementary Course. (*In press.*)

In general method this book is the same with those just described, but, being meant for quite young pupils, it gives more attention to the higher organisms, and to such particulars as can be studied with the naked eye. In everything the aim has been to make clear the cardinal principles of animal life, rather than to fill the pupil's mind with a mass of what may appear to him unrelated facts.

BOTANY. By CHARLES E. BESSEY, Professor in the University of Nebraska.

Advanced Course. Large 12mo. Pp. 611. $2.75.

Aims to lead the student to obtain at first-hand his knowledge of the anatomy and physiology of plants. Accordingly, the presentation of matter is such as to fit the book for constant use in the laboratory, the text supplying the outline sketch which the student is to fill in by the aid of scalpel and microscope.

From J. C. ARTHUR, Editor of *The Botanical Gazette :* "The first botanical text-book issued in America which treats the most important departments of the science with anything like due consideration. This is especially true in reference to the physiology and histology of plants, and also to special morphology. Structural Botany and classification have up to the present time monopolized the field, greatly retarding the diffusion of a more complete knowledge of the science."

Briefer Course. 12mo. Pp. 292. $1.35.

A guide to beginners. Its principles are, that the true aim of botanical study is not so much to seek the family and proper names of specimens as to ascertain the laws of plant structure and plant life; that this can be done only by examining and dissecting the plants themselves; and that it is best to confine the attention to a few leading types, and to take up first the simpler and more easily understood forms, and afterwards those whose structure and functions are more complex. The latest editions of the work contain a chapter on the Gross Anatomy of Flowering Plants.

From J. T. ROTHROCK, *Professor in the University of Pennsylvania :* "There is nothing superficial in it, nothing needless introduced, nothing essential left out. The language is lucid ; and, as the crowning merit of the book, the author has introduced throughout the volume 'Practical Studies,' which direct the student in his effort to see for himself all that the text-book teaches."

CHEMISTRY. By IRA REMSEN, Professor in the Johns Hopkins University.

Briefer Course. 12mo. Pp. 387. $1.40.

An introduction to the study of chemistry, following the inductive method. To avoid overburdening the student's mind, the author has presented a smaller number of facts than is usual in elementary courses in chemistry, but he has at the same time taken pains to select for treatment such substances and such phenomena as seem best suited to give an insight into the *nature of chemical action.* In other words, he has aimed to make the book scientific, to lay stress upon the relations which exist between the phenomena considered, and not to present merely a mass of apparently disconnected facts. Another feature of the work is that principles and laws are treated before the theories which are proposed to account for them.

The other books arranged for in this series are as follows :

PHYSICS. By ARTHUR WRIGHT, Professor in Yale College. (*In preparation.*)

GEOLOGY. By RAPHAEL PUMPELLY, late Professor in Harvard University. (*In preparation.*)

PSYCHOLOGY. By WILLIAM JAMES, Professor in Harvard University. (*In preparation.*)

GOVERNMENT. By EDWIN L. GODKIN, Editor of the *Nation.* (*In preparation.*)

www.ingramcontent.com/pod-product-compliance
Lightning Source LLC
Chambersburg PA
CBHW021514210326
41599CB00012B/1254